An Uneasy Alliance

Professor J. Barkley Rosser (1907–1989)

Director of the Mathematics Research Center, 1963–1973

An Uneasy Alliance

The Mathematics Research Center at the University of Wisconsin, 1956–1987

Jagdish Chandra
The George Washington University
Washington, DC

Stephen M. Robinson
University of Wisconsin, Madison
Madison, Wisconsin

siam
Society for Industrial and Applied Mathematics
Philadelphia

10 9 8 7 6 5 4 3 2

Cover photo courtesy of the University of Wisconsin-Madison Archives, image 13, 029-C1. Used by permission.

Frontispiece photo courtesy of Annetta Hamilton Rosser. Used by permission.

Library of Congress Cataloging-in-Publication Data

Chandra, J.

An uneasy alliance : the Mathematics Research Center at the University of Wisconsin, 1956-1987 / Jagdish Chandra, Stephen M. Robinson

p. cm.

Includes bibliographical references and index.

ISBN 0-89871-535-0 (pbk.)

1. University of Wisconsin—Madison. Mathematics Research Center—History. 2. Mathematics—Research—Wisconsin—Madison—History. 3. Military research—Wisconsin—Madison—History. I. Robinson, Stephen M. II. Title.

QA13.5.W63U553 2004
510'.72'077583—dc22 2003063331

2003063331

We dedicate this monograph

to our wives, Shantha and Chong-Suk,

and

to Vannevar Bush, for his vision and his dedication to U.S. science.

Contents

List of Illustrations

Acknowledgments

The original idea for writing this monograph arose from conversations between Ben Noble and the late Julian Wu. We are indebted to Ben Noble for many valuable exchanges regarding the scope and historical details included in this book. Among the other surviving key members of the Mathematics Research Center, we are particularly grateful to Carl de Boor, Michael Crandall, and Olvi Mangasarian for their insightful comments and input. We are also much indebted to Annetta Hamilton Rosser for her help in many respects, including her comments on drafts.

During the preparation of this monograph, we contacted several members of the scientific community, both within and outside the United States. We received some very thoughtful recollections. For some individuals, we conducted personal interviews. We would to like to acknowledge in particular the following people for their contributions:

Gerald R. Andersen
Kendall E. Atkinson
George E. P. Box
Fred Frishman
James Glimm
T. C. Hu
M. A. Hussain
Joseph W. Jerome
V. Lakshmikantham
Lothrop Mittenthal
M. Zuhair Nashed
Paul Rabinowitz
Klaus Ritter
John V. Ryff
Ram P. Srivastav

Our special thanks and gratitude go to Ivan R. Hershner for his dedication to public service and to MRC. Our conversation with him was insightful and inspiring.

The staff of the Society for Industrial and Applied Mathematics (SIAM) helped us greatly at every step of the production process. Our particular thanks go to Executive Director James Crowley, to Book Operations Manager Alexa Epstein (and later, Mary Rose Muccie), to Developmental Editor Simon Dickey, and to Managing Editor Kelly Thomas for their patient guidance and advice.

During the preparation of this monograph we made extensive use of historical documents in the Archives of the University of Wisconsin-Madison and in the Center for the Mathematical Sciences at the University of Wisconsin-Madison. We are extremely indebted to Sally Ross of the Center for her support and cooperation. The Army Research Office was gracious in providing financial assistance for completion of this monograph through the Scientific Services Program administered by Battelle (Delivery Order 489, Contract DAAH04-96-C-0086). We thank Robert Launer for his encouragement throughout this process. We also convey our appreciation to Kathy Daigle for very efficient work and guidance on contractual matters.

Introduction: Why This Book?

The rounding out of existing theories and the creation of new ones, and the progressive discovery and perfection of mathematical methods, are among the most important desiderata for the future. Their promotion is vital, if the Country's technology is not to be allowed to lapse from a leading position.
—*Army Research and Development,* commenting on the importance of MRC in 1961

What is it about?

This book deals with organizational and scientific issues that arose in the operation of the Mathematics Research Center (MRC), which was operated by the University of Wisconsin (later the University of Wisconsin-Madison) under contract with the United States Army from 1956 to 1987. MRC was one of the first large-scale experiments in government–university cooperation in the United States. The contract was supervised initially by the Office of the Chief of Research and Development (OCRD) and later by the Army Research Office (ARO).

The issues we consider are complex and difficult, involving MRC, its host organization (the Madison campus), and its sponsoring agency (ARO). At the risk of oversimplification, we list four connected issues that recur throughout this monograph. The first two concern the internal structure of a multi-investigator research organization like MRC in a university setting:

- The difficult tradeoff between appointing key scientific staff for long periods (or even permanently) to provide long-term stability and continuity and retaining program flexibility by appointing key staff for shorter periods.
- The question of how much power should reside in the director of such a center and how much should be held by the community of principal research staff, the administration of the host organization, and the sponsoring agency.

The second two issues concern the interaction of a research institute like MRC with its host university and its sponsoring agency:

- The extent to which the culture of the host organization, its rules, personnel practices, and departmental policies can restrict the freedom of action—and even the survival—of an independent center.
- The degree to which a sponsoring agency can itself retain freedom of operation over a lengthy period, as opposed to having to accommodate pressures for short-term results at the cost of de-emphasizing a focus on long-run research.

Throughout this monograph we discuss these issues as they arose at MRC. In the last chapter we distill the lessons learned. For the most part we have left out the routine or easy issues, although many of them were important to the operation of MRC. For example, we say relatively little about the way in which MRC chose the areas to which it would allocate funds for research *except* when those choices were matters of contention or involved external pressures.

It is well known that in August 1970 a terrorist group bombed the building that housed MRC, resulting in the death of a postdoctoral research physicist. Ironically, the victim had no direct association with the Center. This book is not about that bombing; much has been written elsewhere about it, and dealing with it is not within the scope of what we are trying to do. In fact, we say relatively little here about the antiwar movement at Madison, although we do address some of its aspects that help in understanding the organizational climate in which the principal actors had to make decisions.

Who are we, the authors?

We are two applied mathematicians, each of whom was closely involved with both organizational and scientific aspects of MRC over many years.

Jagdish Chandra completed a Ph.D. at Rensselaer Polytechnic Institute with a thesis on nonlinear differential equations. He worked as a research mathematician in an Army laboratory before moving to the U.S. Army Research Office (ARO). He served as a program officer in ARO's Mathematics Division before becoming its director, and for some time chaired the Army Mathematics Steering Committee (AMSC), the organization that provided technical supervision for the Army on the MRC contract. He has now retired from the government and is currently a Research Professor in the School of Engineering and Applied Sciences at the George Washington University and Adjunct Professor of Computer Sciences at the Johns Hopkins University.

Stephen M. Robinson completed a Ph.D. at the University of Wisconsin with a thesis on nonlinear programming, having earlier taken the M.S. at the Courant Institute of Mathematical Sciences at New York University. He served as an administrator, and later assistant director, at MRC and was a member of MRC's principal research staff. He is currently Professor of Industrial Engineering and Computer Sciences at the University of Wisconsin-Madison.

Our combined involvement and time with MRC allowed us to view the important organizational issues that confronted the Center from 1969 until its demise in early 1987, albeit from different vantage points. Because of our different perspectives, we believe that this book offers a unique voice in the discussion of these issues

Why write it, and why now?

We wrote this book to help people who need to deal with issues about the funding of technical research, over an extended period of time, involving multiple actors and sometimes multiple disciplines, and subject to external pressures and demands for accountability. Such issues are particularly salient given the current emphasis on multidisciplinary research, in which managing the generation and transfer of knowledge and technology is a central problem.

The audience that we have in mind is quite broad, including

- managers in government funding agencies, from program officers to agency directors;
- university faculty and administrators, research scientists, department chairs, deans, provosts, chancellors, and presidents;
- research managers in the private nonprofit sector and senior executives of their organizations;
- managers at all levels in for-profit industries that depend on technical research and development. Although some of the issues with which we deal here do not arise in the same form in that sector, others (e.g., perspective, organizational culture, and power sharing) certainly affect for-profits as much as they do nonprofits.

The reason why we wrote this book now is quite simple. MRC was founded in 1956, and it ended in 1987. Many people who played very important roles in its organization and management have already passed on, and others are now at advanced ages. Thus, for the book to use the recollections of living people who had personally dealt with the core issues, it had to be written now.

An alternative, which could be done at any time, would be a book based on the written record. Good records exist, and we have used them freely to supplement the direct testimony of the actors. But we think a work produced *only* from the written record would inevitably have a much different flavor than the one we have written.

Why read it?

As we stated earlier, MRC was one of the first large-scale experiments in government–university cooperation in the United States. Among other academic research centers in mathematics, only the Courant Institute of Mathematical Sciences goes as far back in time, and the organizational structure of Courant is somewhat different from that adopted at MRC. Thus, many of the issues recounted here arose for the first time at MRC, and the people who faced them had to find new solutions. Some of what they tried worked very well; some did not. Knowing which solutions worked and which did not should help people who have to make such decisions now.

This is not a purely historical account of what happened at MRC. Rather, we use history as a vehicle to draw some conclusions that we hope readers will find useful. Although many of these issues arose for the first time at MRC, they have since emerged in other places and will inevitably recur in the future. People who must manage technical research will face these issues again and again, in different contexts. We hope that this book will help them to understand these issues and to find their own solutions.

How is it organized?

We have organized this book in six main chapters, plus appendices. Here we give a brief preview of the topics covered in each of these.

Chapter 1 starts with some brief general observations about science and its role in society, then turns to more concrete considerations of the relationship between science and government in the United States, especially in the 20th century. It introduces the U.S. Army's use of science and mathematics, and from this proceeds to describe the foundation of MRC in the mid-1950s and its early operation under Rudolph Langer, the first director. The latter part of this chapter lays the groundwork for future chapters in examining the organizational climate of the University of Wisconsin at Madison, the site of MRC, and the relationships between MRC and other units on the campus. This information is important for understanding the developments described in later chapters.

Chapter 2 describes the way in which MRC operated under the second director, J. Barkley Rosser, from 1963 to 1973. This was the time during which the Center had the greatest flexibility of operation and the greatest freedom to define its own research program. We look at how Rosser and the permanent staff shaped that program and how they struck a balance between broadly based and more focused research. In the latter part of the chapter we see how MRC, along with many other U.S. institutions, was caught up in the political turmoil centered on the Vietnam War. The effects of that turmoil were to change the Center's environment and its relations both to the rest of the Madison campus and to the Army. Those changes, combined with the general decrease in support for mathematical research that we also describe in this chapter, were eventually to destroy the Center.

The third chapter describes the 14-year period from the end of Rosser's directorship to the end of MRC itself. We show how the pressures described above led to substantial changes in the administrative operation of the Center. The institution of "permanent members," which had earlier brought benefits in focus and continuity to the research program, now seemed too rigid and inflexible to allow adaptation to the new environment. Accordingly, the terms of appointments changed, and at the same time the Army experimented with new oversight mechanisms for the Center, in particular instituting a series of peer reviews through which it could exert pressure for changes in research directions. Neither of these was ultimately enough to retain the Center's funding, and the chapter ends with the demise of MRC.

Thus, the first three chapters are organized chronologically, though their purpose is not a chronology per se but rather a view of the dynamic evolution of the Center's research program in response to the initiatives of the participants and the external pressures imposed by political events and by organizational changes in the funding agency. The last three chapters take a different viewpoint, trying to assess the impact of MRC on people and on issues in applied mathematics.

Chapter 4 looks at several areas of MRC's research and tries to convey something of the flavor of this work and the impact it has since had. The most famous of these is probably the work on spline functions, but viscosity solutions may be a close second.

Chapter 5 examines the human side of MRC, looking at its impact on the many visitors as seen through the recollections of some of these. It also includes a very incisive analysis of MRC's strengths and weaknesses by an external reviewer, James Glimm.

Chapter 6 tries to distill from the rest of the account some lessons that MRC's experience has taught us. We look at its role as a pioneer in center-based mathematical research and compare it with four other research centers. We then discuss the complex relationships among the funding agency, the research center, and the nature of the discipline that ultimately shaped MRC's accomplishments, and we consider how MRC's experience in turn influenced the shape of future initiatives in Department of Defense–funded research. Finally, we return to the four recurring issues with which we began the book (in the section on "What is it about?" above) and ask what light MRC's experience can shed on them.

The two appendices present additional views on the relationship between universities and government-funded research (particularly defense research), in a form particularly relevant to our discussion of MRC. They were written originally as memoranda, the first by Stephen C. Kleene (who served as acting director of MRC and later as Dean of the College of Letters and Science at Madison) and the second by J. Barkley Rosser. The book also contains a selection of photographs of some of the people whose activities appear in these pages.

Finally, packaged with the book is a CD containing two files that list all of MRC's publications. The two files together cover the 30-year period from March 1957 through February 1987, with a slight overlap: the first file covers March 1957 through June 1976, while the second covers July 1975 through February 1987. The files are in PDF format and are searchable with the Adobe® Acrobat® viewer; in addition, they contain indices by author and by AMS subject classification. Although the files originally existed electronically, in ASCII text format, the authors were unable to find copies of those; accordingly, the files given here were produced by scanning printouts that were still in existence.

The nearly three thousand technical summary reports (TSR) that MRC produced constitute the majority of the publications listed in these files. Listings for these reports include the National Technical Information Service (NTIS) accession number: e.g., AD147634 for TSR 1. Journal publications resulting from reports are also noted if available at the time the index was prepared. The files also list fifty-four commercially published books resulting from advanced seminars and symposia that MRC organized. These reports and their published versions, together with the books, constitute the written portion of MRC's scientific legacy. The unwritten portion comprises the difference that MRC made in the scientific careers of its many staff members and of their students, and in the practice of applied mathematics.

Chapter 1: The Beginning, 1956–1963

This decision brings to us an opportunity and a responsibility, an opportunity to make Wisconsin one of the great mathematical centers of America, a responsibility to do so in the interest of our national strength both military and scientific.

Edwin B. Fred, President, The University of Wisconsin, Madison, 1955

Introduction

This chapter describes the beginning of MRC and tries to place it in the context of governmental efforts to exploit the discoveries of science for the public good. We start with a brief résumé of interactions between science and society extending back to ancient times. Then we describe the development of cooperation between government and science in 20th-century America to the end of World War II. At mid-century, several governmental science agencies were already in place and functioning, one of which would found and support MRC. We describe the relationships at that time between mathematics and the military, and more particularly between mathematics and the U.S. Army, leading up to the founding of MRC in the mid-1950s. We conclude with a picture of MRC's initial operation, its early impact, and its relationships both with the university in which it existed and with the Army that supported it. Even at this early stage these relationships brought out some of the key issues that would recur throughout MRC's life.

Science and society

For many centuries science has played a role in shaping society. Ancient Chinese, Egyptian, and Indian societies offer many examples of how the ruling elite sought and used the counsel of scientists and philosophers. In Western civilization, Hero, King of Syracuse, asked Archimedes to devise means for fortifying and defending his city. After Hero's death these defenses successfully withstood Roman attacks from land and sea and forced the Romans to resort to a blockade.

Politicians, statesmen, and military leaders have long consulted and sought assistance from scientists for solutions to problems of war and peace. History offers many instances in which such nonmilitary specialists have been able to assist the military with new ideas about the machines, the tactics, and the strategies of war.

From the beginning of the 20th century there have been specific instances in this country of significant relationships between government and science. For example, as early as 1900, the U.S. Army instituted a program to seek a means of removing or preventing excessive weed growths in waterways. They sought assistance from university scientists on an aquatic plant control program. However, the history of the science–government relationship in the United States before the start of World War I was one of benign aloofness. Scientists feared governmental interference in their activities and were reluctant to receive any financial assistance from government on the grounds that it might jeopardize their intellectual freedom. The government, on the other hand, was generally committed to the liberal democratic tradition and so hesitated to intervene in activities that traditionally had been left to market forces.

Some 70 years after the establishment of the National Academy of Sciences (NAS) in 1863, an editorial in *Science* (the journal of the American Association for the Advancement of Science) commented:

> The liberal spirit which animates both Congress and executive departments in their dealings with scientific affairs is very apt to lead them into the support of scientific enterprises without any sufficient consideration of the conditions of success, and of efficient and economical administration, and careful consideration of each proposed undertaking by a committee of experts is what is wanted to insure the adoption of the best methods.

The World War I began a reassessment of the traditional hands-off attitude toward science and, as we show in the next section, World War II changed the environment completely.

Government and science in 20th-century America

Between 1900 and 1940 the U.S. population increased from 75 million to about 130 million. In some countries, comparable increases have been accompanied by famine. In the United States, this population growth corresponded with a more abundant food supply, better living, more leisure time, and better health. One can argue that this prosperity was largely due to a combination of three factors: the free play of initiatives of a vigorous people under democracy, the abundance of natural resources, and advances in science and its applications. Without scientific progress, no amount of achievement in other directions can ensure a nation's health, prosperity, and security in the modern world. However, science by itself provides relief from no individual, social, or economic ills. Whether in peace or war, science is a necessary but not a sufficient condition for advancing national welfare.

While several isolated instances illustrate the increasing relationship between government and the scientific community both during and after World War I, the really significant change came as a result of the experience of World War II. In the former case, efforts were initiated in the area of aeronautics. During the Great Depression, as a result of the joint efforts of leaders of the scientific community within the Academy of Sciences and of the Secretary of Agriculture, a Science Advisory Board was created by executive order to advise the President. However, the board's attempts to establish a basic research program in universities did not succeed.

The most significant step toward a durable relationship between government and science came in 1940. The war raging in Europe presented an opportunity for scientific work to affect a conflict. The leaders of the scientific community began to lobby for the creation of a government agency that would mobilize U.S. scientists for the country's inevitable entry into the war. As a result, President Roosevelt created the National Defense Research Committee (NDRC) under the chairmanship of Dr. Vannevar Bush. Bush was a former Dean of Engineering at MIT and was later the president of the Carnegie Institution in Washington. NDRC was the forerunner and foundation of one of the strongest scientific research and development establishments that this nation and the world have ever witnessed. Members of the NDRC were drawn from the universities, government departments and agencies, private foundations, and industry. The mandate of this group included the correlation and support of scientific research on the mechanisms and devices of warfare, with the exception initially of those relating to war in the air.

A major landmark in the progress of governmental support for science in the United States turned out to be the creation of the expanded Office of Scientific Research and Development (OSRD), under Vannevar Bush. This initiated a structure under which U.S. scientists were brought into war efforts through a contract mechanism, while leaving them free to pursue their creative work. The contractor (investigator) simply agreed to carry out studies or experimental work in connection with a given problem and to report to the customer, or OSRD. No attempt was made to dictate the manner of investigation, nor were the contractors compelled in any way to take on the tasks. Through the OSRD, for the first time in U.S. history, substantial government funding was channeled to the universities, either directly or by the establishment of associated laboratories. Examples of the latter include the Radiation Laboratory (MIT) and the Los Alamos Atomic Weapons Laboratory (University of California).

There is no doubt that OSRD was one of the most significant milestones in the development of government–scientist relations in the United States. Towards the end of the war, President Roosevelt asked Bush how the OSRD experience in wartime could be utilized after the war. Bush responded with a seminal report, *Science, the Endless Frontier,*[1] from which we briefly highlight some major points:

- The publicly and privately supported colleges, universities, and research institutions are the centers of basic research. They are the wellspring of knowledge and understanding. As long as they are vigorous and healthy and their scientists are free to pursue the truth wherever it may lead, there will be a flow of new scientific knowledge to those who can apply this to practical problems in government, industry, or elsewhere.

- Many of the lessons learned in wartime applications of science under government can by profitably applied in peace.
- Scientific programs on broad fronts result from the free play of intellects, working on subjects of their own choice, in the manner dictated by their curiosity for exploration of the unknown.
- In World War II, it has become clear beyond all doubt that scientific research is absolutely essential to national security.
- Our ability to overcome possible future enemies depends upon scientific advances, which will proceed more rapidly with diffusion of knowledge than under a policy of continued restriction of knowledge now in our possession.

In succeeding chapters, as we watch the evolution of MRC at the University of Wisconsin, the significance and implications of these comments will become evident.

Government research agencies at mid-century

In 1946, the Office of Naval Research (ONR) was created to plan, foster, and encourage scientific research and to provide within the Department of the Navy a single office which by contract or otherwise was able to sponsor, obtain, and coordinate innovative research of general interest to all sectors of the Navy. The individuals responsible for the successful operation of ONR firmly believed that the primary aim of much of the research that it funded should be free rather than directed. That is, it should explore and understand the laws of nature and not necessarily be aimed at solving practical problems. By and large, the naval authorities believed that most of the basic research carried out under ONR's auspices should be published in the normal way. This policy allayed many fears in the academic and scientific community. The office began to take on the role that was envisaged for the yet-to-be-established National Science Foundation (NSF).

The National Institutes of Health (NIH), established in 1930 and generously funded by OSRD during the war, became a major focus of government support for medical research in the universities. The Atomic Energy Commission (AEC) was established in 1946, and this agency forged close links with universities by contracting research work to them and by building up the university-associated laboratories that it had inherited from the Manhattan Project. These laboratories have become well known as centers of basic and applied scientific research.

When the NSF was eventually established in 1950, defense research was excluded from its terms of reference. In the initial recommendation, Dr. Bush had envisioned defense research as one of the organizational component of NSF's charter. As a consequence, the Department of the Army, and subsequently the Air Force, established their own offices of research. The Department of the Army's Office of Ordnance Research was established in June 1951 on the campus of Duke University.

Mathematics and the military

The military has long used the talents of civilian mathematicians, both in the United States and in other countries. One can document the history of this association for several decades, perhaps even for centuries, but it is particularly evident during and immediately after World War II.

World War I had marked several examples of early operations research on both sides of the Atlantic in the efforts to formulate and analyze military operations mathematically. In England, F. W. Lanchester modeled complex military operations with ordinary differential equations. While these efforts did not have any appreciable effect on the operations of World War I, they were forerunners of later combat models. In the United States, Thomas A. Edison made studies of antisubmarine warfare for the Navy. His work included the compilation of statistics to be used in determining the best methods for evading and for destroying submarines, the use of a "tactical game board" (compare this with recent electronic sand tables) for solving problems of avoiding submarine attack, and an analysis of the value of zigzagging as a method for protecting merchant shipping.

Operations research began as an *organized* form of research directly before the outbreak of World War II. An interesting example of operational analysis from early in the war is the famous "Blackett's Circus." During World War II, the British Army had major responsibility for air defense. In 1940, General Pile, commander in chief of the Anti-Aircraft Command, sought scientific assistance in coordinating the radar equipment at gun sites. Professor P. M. S. Blackett of the University of Manchester, a fellow of the Royal Society and Nobel laureate, led a team of mathematicians, physicists, physiologists, and army officers in studying this problem. "Blackett's Circus," as the group was called, were able to demonstrate the value of the mixed-team approach to such operational problems.

Another instance of the contribution of mathematics to war efforts in World War II is the work of the British mathematician Alan Turing in breaking the German Enigma cipher. One of the key reasons for the high efficiency (and resultant lethality) of the German armed forces was their extensive use of radio communications. Messages sent in this way had to be encrypted, and the encryption system was purposefully difficult to break. Turing headed a very talented group of British scientists at the British Communication Headquarters, Bletchley Park.

Turing was more than just a brilliant mathematician and scientist. He was also a philosopher, an unconventional genius, and a man ahead of his time. After the war, he worked on theoretical aspects of computer design until his untimely death. Today, he is credited with being a founder of computer science and artificial intelligence. He spent considerable time at Princeton, where he was exposed to a galaxy of mathematicians and physicists at the Institute for Advanced Study (IAS). These included legendary figures such as Einstein, von Neumann, Courant, Gödel, and Lefschetz. Professors Richard Courant and Solomon Lefschetz were credited with establishing two very visionary mathematical activities, namely, the Courant Institute of Mathematical Sciences and the Research Institute for Advanced Studies. The former is a premier institute in mathematical sciences that had considerable initial impetus from ONR and AEC, whereas the latter had significant support from the Air Force. However, Turing's interest in logic also brought him in contact with Stephen C. Kleene and J. Barkley Rosser, who were later to play very significant roles in the evolution and consolidation of MRC.

There was a conscious effort on the part of several U.S. mathematicians to ensure that a substantial number of mathematicians were available for war efforts. Some of the key leaders in this effort were Dr. Mina Rees (who had a long association with ONR), Professor Oscar Veblen, and Lee Zippin, an army corporal. Nowhere was this effort more visible than at the Ballistics Research Laboratory (BRL) located in Aberdeen, Maryland. Zippin, in particular, followed the development of the first electronic computer, the ENIAC, at the Moore School of the University of Pennsylvania. ENIAC was eventually moved to BRL. Professor von Neumann and his associate, Dr. Herman Goldstine, played a key role in the development of this first major digital computing platform. It is interesting to observe here that Goldstine continued his association in an advisory capacity with MRC until the mid-1970s.

Outsiders often consider mathematics to be a tool or a framework. Rarely do people expect new technical developments to arise from work in mathematics. This attitude seems to be seriously incomplete, as shown by current developments in computer science and by the synthesis of new materials or the analysis of structural patterns in biological systems. Nonetheless, during this period work on concrete problems such as ballistic trajectory analysis, firing tables, digital computing, data analysis, and cost-effective and efficient day-to-day operations influenced many mathematicians in their future work.

One individual in this group who played a key role in the early life of MRC was Professor I. J. Schoenberg, who joined BRL in 1943 to work on ballistic trajectories. As Schoenberg has described,[2] people then did these calculations by hand or with desk calculators. The numerical integrations, which took into account a multitude of drag functions for air resistance obtained by firing various projectiles, were quite tedious. With ENIAC, one could perform these calculations more rapidly and with smaller integration steps. In these methods, however, the accumulation of round-off errors, due to the rough drag function tables, was unacceptable; they needed to be smoothed by being approximated by analytic functions. This became the central problem of interest to Schoenberg, which he solved (and for which he developed a comprehensive theory) by what he called (in subsequent work at MRC) *cardinal spline interpolation* and *cardinal spline smoothing*. Schoenberg, along with Professor Carl de Boor and several of their associates, was later responsible for the flourishing work at MRC in approximation theory using splines. We will return to this topic in Chapter 4.

Mathematicians played an important part in civilian research labs during World War II. Although they often possessed little knowledge of the technical subjects they were assigned to work on, they were quicker to learn new topics that used mathematics than were those without a mathematical background. Rosser's work presents a good example of this phenomenon. His initial training was in abstract logic. Early in his career, during World War II, he became involved in ballistics. After the war, he continued to be involved in government think tanks such as the Institute for Defense Analyses (IDA), while vigorously pursuing his mathematical career. IDA was founded in 1947, when Secretary of Defense James Forrestal established the Weapons Systems Evaluation Group (WSEG) to provide technical analyses of weapons systems and programs. In the mid-1950s, the Secretary of Defense and the Chairman of the Joint Chiefs of Staff asked MIT to form a civilian, nonprofit research institute. The institute would operate under the auspices of a university consortium to attract highly qualified scientists to assist WSEG in addressing the nation's most challenging security problems. It is interesting to note that in 1958, at the request of the Secretary of Defense, IDA established a division to support the newly created Advanced Research Projects Agency (ARPA). Rosser brought to his duties as MRC's second director an enlightened perspective on how mathematics and science research can benefit from contact with practical problems, while at the same time ensuring advances in mathematics that benefit the society both in peace and in conflict.

The Army and mathematics

Army general officers such as Lt. Gen. Arthur Trudeau, Lt. Gen. James M. Gavin, and Brig. Gen. Chester Clark, and other officers such as Lt. Col. Ivan R. Hershner, recognized early in the 1950s that the Army is a major user of the fruits of research in mathematics, no matter what the source is. The recent experience of World War II further confirmed this belief for these individuals, and they thought the Army had an obligation to plant some seeds to replenish this source of knowledge. Although this belief was not widely held within the Army, these enlightened officers and other members of the Army establishment were successful in convincing the Army to establish a center of mathematical expertise at an academic institution.

In preparation for this crucial decision, the Mathematics Advisory Panel of the Army, a precursor group to the Army Mathematics Advisory Group (AMAG) and the Army Mathematics Steering Committee (AMSC), conducted a survey of the uses of mathematics in Army activities and combined that with a census of its mathematically trained personnel and its expenditures for mathematical investigation. It attempted to pinpoint, in some qualitative way, where the lack or insufficiency of existing mathematical theory might be restrictive, or where mathematical difficulties were critically impeding technological progress. The motivation for such a survey was primarily rooted in the concerns of cold war realities. The principal opponent, the Soviet Union, was correctly perceived to be quite strong in the fields of mathematics and related sciences. Due to the technological progress of the 1920s and 1930s, World War II was fought by means quite different from those employed in World War I. Moreover, if there were to be another war in the 1950s, its prosecution would depend on means that were quite unlike those of World War II. Thus, there was a constant evolution of weapons, tactics, and strategies of warfare.

From the analysis of this survey, it was not very difficult for the Advisory Panel to conclude that the Army was not utilizing all of the advantages that mathematics could offer. Therefore, it seemed of the greatest importance from the standpoint of national security for the Army and the country to remain at the forefront of mathematical developments. The Advisory Panel made two recommendations. First, it advised that the Army establish for itself a mathematics research center at an academic institution. The key aspects of the work statement were to conduct basic research in selected areas of mathematics relevant to the interests of the Army, to provide educational and training courses to the Army on current mathematical methods, and to be available for consulting on mathematical problems encountered by Army scientists and engineers. It was to carry on research in four areas, and we describe these here as they formed the basic framework for MRC's activity throughout its lifetime.

- *Numerical analysis.* This was broadly understood as the adaptation of mathematics to high-speed computation, to include the use of electronic computing machines, the formulation of mathematical problems for exploration by such computers, and hence the broadening of the field in which such computers could be used. This area was originally intended to include "the engineering physics of high-speed computers," presumably what is now referred to as computer architecture and computer engineering, though unfortunately very little was in fact done at MRC in those areas.
- *Statistics and the theory of probability.*
- *Applied mathematics,* including ordinary and partial differential equations as well as *physical mathematics* with emphasis on fluid mechanics, elasticity, plasticity, electrodynamics, electrical networks, wave guidance, and propagation.
- *Operations research,* including such subfields as linear and nonlinear programming, game theory, decision theory, information theory, and optimization.

Second, the Advisory Panel recommended that it be recognized and established as a continuing body, with the assignment to inform itself about new mathematical developments and to keep itself informed of the Army's needs in and uses of mathematics, to supervise activities of this kind, and to facilitate the interchange of relevant information between activities. Initially, this was a committee of about twenty-five, including four from academic institutions. The rest represented various Army activities.

Towards the realization of the first recommendation, the chief of research and development of the Army appointed Ivan R. Hershner (then the chair of the Mathematics Department at the University of Vermont) to head an effort to explore with various universities and research groups their possible interest in this center. Letters were sent to over fifty U.S. institutions of higher learning. Based on the level of interest expressed, this small group of experts visited twenty-six universities in a span of six weeks, meeting with both faculty and senior university administrators. Universities included some of the leading graduate schools such as Brown, Columbia, the University of Chicago, Duke, California Institute of Technology, Harvard, the University of Illinois, the University of Michigan, MIT, New York University, the University of North Carolina, UCLA, UC-Berkeley, Stanford, the University of Wisconsin, and the University of Virginia.

This process resulted in twenty-one formal proposals. A technical advisory committee of Army scientists evaluated these proposals against criteria such as the quality of available faculty, prospects for strong technical leadership, suitability of the facilities, and the readiness of the university to undertake such a responsibility. The Army had offered to provide a state-of-the-art computer, but it expected that the selected university would supply suitable physical space to house the center. Based on these criteria, the review team recommended the University of Wisconsin at Madison as the outstanding choice.

Establishment of the center

The decision to establish the Mathematics Research Center at the University of Wisconsin was announced on November 16, 1955, by Lt. Gen. James M. Gavin, chief of research and development of the U.S. Army. The university community at Wisconsin welcomed this announcement. As President Fred of the Madison campus stated:

> This decision brings to us an opportunity and a responsibility, an opportunity to make Wisconsin one of the great mathematical centers of America, a responsibility to do so in the interest of our national strength both military and scientific.

This announcement by Gavin was the culmination of a long process of study by the Army Advisory Panel of the interests and requirements of the Army's technical community, the solicitation of proposals from different academic institutions, and the subsequent evaluation of these proposals against set criteria such as scientific excellence, potential relevance of proposed areas of research, overall quality of the institution and its location, and its willingness to enter into partnership with the Army in providing technical assistance. Under all of these criteria, the University of Wisconsin was rated quite high and Madison was judged as an excellent place to live. The institution was also identified as a premier mathematical resource.

The Army described the general goal of the center as follows:

> to provide a nucleus of highly qualified mathematicians who will carry on investigations in mathematics of interest to the Army and who can be called upon for advice on specific problems beyond the capability of Army facilities.

In addition, the Army took a rather broad perspective in believing that in addition to fulfilling an Army need, the center would aid the national effort in mathematics research and would increase the availability of trained mathematicians.

The first contract for MRC's operation was signed on April 25, 1956,[3] and the university designated Professor Rudolph E. Langer as MRC's first director. Langer was a long-standing member of the Department of Mathematics. A Harvard graduate (see biographical sketch below), he had taught at Dartmouth College and Brown University before coming to Wisconsin. He had a broad technical background with specialization in the theory of ordinary differential equations, and was considered a leader in the mathematical community.

The greatest challenge that Langer faced was how to attract worldwide talent to the center. The structure of MRC consisted of a core of permanent members who generally had (at least initially) a tenured appointment in the Department of Mathematics. These were high-ranking experts in several fields corresponding to the statement of work described earlier.

The next tier of scientific staff consisted of two groups. In the first group, several eminent mathematicians of the country and of the world were invited to spend shorter or longer periods of time at the center as visiting scholars. This group included some individuals from the Madison campus and some who were on leave of absence from their home institutions or departments. Through this gradually changing group, the center kept itself in fresh and up-to-date contact with global mathematical developments. It was deliberately and perpetually stimulated by the impact of different points of view and personalities.

The second group consisted of a corps of younger junior faculty or postdoctoral staff members, typically at the ranks of assistant professor or research associate. While the members of this second group benefited immensely from the associations with persons of more mature accomplishments, they also brought new ideas and vigor to the center.

Professor Rudolph E. Langer

Rudolph E. Langer, the first director of the Mathematics Research Center (MRC), was born in Boston on March 8, 1894. He received an S.B. (magna cum laude, 1918), an A.M. (1920), and a Ph.D. in mathematics (1922) from Harvard University. After three years as an instructor and assistant professor at Dartmouth College and a year as an assistant professor at Brown University, he joined the faculty at the University of Wisconsin as a professor of mathematics. He retired from this position in 1964 after thirty-seven years of distinguished service.

During World War I, Langer was employed in confidential government services in New York City. He had visiting professorships at Göttingen, Ohio State University, Harvard, Stanford, and the University of Texas. He served as chairman of the Department of Mathematics from 1942 to 1951 and as the chairman of the Division of Humanities for a couple of years. In 1956, Langer was named the first director of MRC. He served as the MRC director until his retirement in 1964. Under his direction, the center evolved into a major international research force. More than fifty research mathematicians hailing from the United States, Europe, South America, and Asia staffed MRC. In recognition of this service to the nation, Langer was awarded the Outstanding Civilian Service Medal from the U.S. Army.

In his capacity as MRC director, he served as a worldwide ambassador for the university. He was the leading architect in the development of mathematics at the University of Wisconsin. He guided over twenty students to their doctoral degrees, and many of his students became professors at prominent U.S. universities. Langer made significant contributions to the fields of differential equations and asymptotic theory and published over fifty papers. He also served long terms as an editor of the *Bulletin of the American Mathematical Society* and the *Duke Mathematical Journal*. Langer held other distinguished professional offices, including president of the Mathematical Association of America, vice president of the American Mathematical Society, and vice president of Section A of the American Association for the Advancement of Science.

In addition to his interest in mathematics, Langer had a passion for art. He and his wife, Louise Stemler, owned a collection of prints that spanned the history of graphics from Dürer to Matisse. Professor Langer died on March 11, 1968. At the time of his death, their collection of 300 years of Japanese prints was on display in the Madison Art Center.

Langer was a great believer in the role of strong leaders in the society. In a 1959 address to a group at the University of Michigan, he stated, "only the shrewdest use of our wits can win the Cold War." He continued:

"In any society, the genius for leadership resides in only a small minority of the people. But the whole impetus towards advance, the whole implementation of progress, comes from this few. In statesmanship, in science, in education, in the arts, and in business, the leaders are the exception. The great majority of any population plays no role in shaping its own destiny, but it depends inertly upon the leaders, who are the brain of the society body. The best that education can do for the many average pupils would come to little were they not to receive guidance later from the superior minds."

Early impact

One of the important features of the scientific program at the center was its emphasis on computing. Early in its existence, MRC acquired a large computer as one of its research instruments, and this helped it to have a significant impact and influence on the establishment of a genuine applied mathematics program. During the recent World War II years, a considerable impetus had been given to applied mathematics in general, and more specifically to some topics of mathematical physics, including continuum dynamics, classical electrodynamics, theories of elasticity and plasticity, and fluid mechanics. Certainly these topics were reflected in the scope of work of the center from the beginning. Partly due to wartime experience and partly due to the natural outgrowth of normal industrial development, there was an increasing reliance on automation and control procedures. These methods naturally generated needs for automatic electromechanical or electronic computing. Other technical advances that provided evidence for similar needs were modern radar, fire control, and television techniques. These technological phenomena pointed naturally toward the requirement for large-scale, high-speed automatic computing and for the design of computing machines that would offer such performance.

Two directions of inquiry motivated the program in numerical analysis: the identification of mathematical needs for high-speed, automatic computing, and the identification of characteristics of computing devices that would be effective in various phases of mathematical analysis. Early on, the center assembled experts and organized symposia that effectively addressed some of the relevant technical issues. Many promising technical trends started to emerge. These included the numerical solution of problems in continuum physics, approximation and spline theory, and more generally, a sound framework based on applied functional analysis for convergence and error analysis of numerical procedures.

Experiences of war also pointed to the significance of statistical methods and of the problems those methods could address. There was then no single statistics department at the University of Wisconsin, but there was a "Division of Statistics," comprising members from all over the campus, including departments from agriculture, mathematics, and engineering. MRC was very influential in the establishment of a Department of Statistics with a focus on statistical methods. This department was able to provide intellectual leadership and support to users in business, economics, engineering, agriculture, and the medical school. This department also proved quite beneficial to the Army, as its expertise was called upon frequently by various Army organizations. The statistical aspect of MRC's activity also played a substantial role in educational programs of short courses and tutorials for Army scientists and engineers.

Another significant impact of the establishment of the center was the creation of an excellent environment for students and young researchers. This intellectually rich milieu included visitors from other U.S. and foreign institutions. A unique feature of MRC was that it encouraged a variety of University of Wisconsin faculty and visitors with broad interests to interact with each other.

As we will outline in the rest of this monograph, the greatest legacy of MRC is that it and the Courant Institute of Mathematical Sciences led the mathematical sciences community in the development of a research agenda and in defining directions for research in applied mathematics. At least initially, MRC did not have to focus its research activities in a way that would seriously restrict or channel their development. However, as resources became limited, this had to change. This aspect of the center will be explored in Chapter 3.

MRC provided a balanced view of mathematics, in contrast to other trends of the 1950s and 1960s that emphasized the separation of "pure" and "applied" mathematics. There was a sharp dichotomy in the mathematical curriculum between these two areas. MRC was well disposed towards the development of a genuine applied mathematical program that was stimulated by practical problems but that also encouraged intellectual curiosity and was not constrained by the limitations of any specific applications. MRC emphasized the need for mathematics to return to reality. John von Neumann, a great contributor to both pure and applied mathematics, put it this way:

> As a mathematical discipline travels far from its empirical source, or still more, if it is a second and third generation only indirectly inspired by ideas from "reality," it is beset with very grave dangers. It becomes more and more purely aestheticizing, more and more purely *l'art pour l'art.* This need not be bad, if the field is surrounded by correlated subjects, which still have closer empirical connections, or if the discipline is under the influence of men with an exceptionally well-developed taste. But there is a grave danger that the subject will develop along the line of least resistance, that the stream, so far from its source, will separate into a multitude of insignificant branches, and that the discipline will become a disorganized mass of details and complexities . . . whenever this stage is reached. The only remedy seems to me to be the rejuvenating return to the source: the reinjection of more or less directly empirical ideas. I am convinced that this was a necessary condition to conserve the freshness and vitality of the subject and that this will remain equally true in the future.[4]

MRC was true to these principles from the start and, as we will see in the next two chapters, the intellectual leaders of the center went far toward making MRC a great center for applied mathematics. The association with the Army also provided a constant source of challenging practical problems.

As is well known, there was a serious shortage of mathematically trained persons during that time throughout the United States. This shortage was felt by the Army no less than by other government agencies, by industry, and by the universities. Therefore, the center became an important source for training young scientists in mathematical sciences. This, of course, was done in cooperation with other departments, particularly with mathematics departments, at Madison and elsewhere through pre- and postdoctoral training support. The Army was also encouraged to send persons for various durations of further in-service training with the experts at the center, to enhance their educational level in mathematics or to broaden the scope of their effectiveness in the use of mathematical techniques.

Secrecy and security clearance at MRC

Institutions of higher learning have always cherished ideals of creating and disseminating knowledge in an open environment. Intellectual pursuits flourish only in such environments, where there is a free flow of information and where unfettered collaboration can take place. One major contributor to this flow of information is publication in the open research literature. The basic contract between the Army and the University of Wisconsin recognized this fact at the very outset. Publication in recognized scientific journals of papers resulting from research performed by the center personnel was highly encouraged. In fact, this was considered as a key metric for the impact of the center.

A critical objective of the center was to provide advice to the Army and assistance on mathematical techniques. The center offered stimulating scientific contact and cooperation between its key staff and Army scientific personnel. These interactions provided a significant source of challenging technical problems. Those problems, in turn, encouraged novel mathematical techniques and stimulated the development of a genuine applied mathematics program. In order to facilitate meaningful interactions with various Army agencies, it was necessary that some key members (primarily the permanent members) possess security clearance. While no classified work was carried out at the center, possession of a security clearance facilitated interaction between senior members of the center and scientists at Army installations.

With this practical goal in mind, the director of the center always encouraged the permanent members to have security clearance. It should be pointed out that there was no contractual requirement that the key staff of the center have security clearance. Initially, this did not pose a major problem, as many of the permanent members hired in early years had had experience in working with national security agencies, especially during the war years. However, it became a growing source of concern on the campus during subsequent years. As we will see in the next chapter, the topics of secrecy, security clearance, and consulting with the Army would create pretexts for activism and agitation in an already charged campus atmosphere.

Relationship with mathematics and other departments

At the University of Wisconsin at Madison, the mathematics department and the majority of the faculty were initially supportive of MRC. The department viewed MRC as beneficial in providing them with extraordinary opportunities for expansion. After all, the first director of the center was a long-standing faculty member of the department. However, this was not to be the case in subsequent years. An exploration of the dynamics of the MRC–department relationship is worthwhile.

While "mathematics" was part of the title of the center, the focus of MRC was primarily related to research on applied mathematics and its applications. Several members of the mathematics department considered themselves to be applied mathematicians; however, their activity covered only a part of what the center planned to accomplish. Moreover, the center operated as an autonomous institute and had the potential for collaboration with more than one department.

Another source of concern was the salary structure at the center compared to that of the mathematics department. In order to attract the best mathematicians, the center offered salaries to key members that were above the average salaries in the mathematics department. These and other perceived differences aggravated an already tense atmosphere and promoted misunderstandings during the subsequent campus unrest. We explore this topic further in Chapter 2.

MRC was not an academic department, so it could not award tenure, for which it had to depend on one of the university departments. Mathematics and several other departments therefore stood to benefit considerably from the resources of MRC. We discuss the evolution of these relationships throughout the rest of this monograph.

MRC and the university

As we described earlier, the University of Wisconsin's administration was quite supportive during the establishment of the center. MRC continued to enjoy considerable support at the senior level of the university administration, particularly from the deans of the College of Letters and Science and the College of Engineering, and from the president (later chancellor). While there was initially no permanent advisory committee at the university level for MRC, such a committee did evolve in subsequent years, as we describe below. This University of Wisconsin Advisory Committee, appointed by the chancellor, provided advice both to the chancellor and to the director of MRC on matters concerning the relationship of MRC to the university and its various colleges and departments. It had no executive power, but its advisory role strengthened the hand of the center director in matters of joint appointments with departments and in the planning of programs that involved the cooperation of more than one college or department.

With the increasing establishment of multidisciplinary research centers, the experience of MRC in this matter could be helpful. The director of MRC was confronted with a difficult balancing act. Departments believed that MRC existed for their benefit. The director, on the other hand, had to reconcile this view with the Army contract and to interpret the role of MRC in that light. The advisory role of senior university management can provide considerable protection to a center director against possible misunderstandings, either with departments (e.g., in appointments) or with the sponsoring agency (e.g., in issues such as security clearance).

MRC and the Army

In response to the second recommendation of the Army Advisory Group, the chief of research and development of the U.S. Army (CRDA) established an Army Mathematics Steering Group (AMSG), which was later renamed the Army Mathematics Steering Committee (AMSC). This committee served as a liaison between the center and Army activities, facilitating interactions between the center staff and Army scientists and engineers. The committee, which reported to CRDA periodically, reviewed the technical work of the center, disseminating technical information to relevant Army groups. At the same time, it apprised the staff of the center of some outstanding technical problems potentially amenable to mathematical modeling, analysis, and solution. This group also facilitated education programs of short courses and tutorials presented by the center staff at selected Army sites.

The Army has many pressing mathematical problems, but MRC generally was not a place where the Army could take its urgent and short-term problems for quick solution. MRC was well situated, however, to advise on longer-term technical challenges. There was an increasingly vocal element in the Army whose expectations about MRC included solution of immediate problems. Some of these elements felt that the money spent on MRC could be better used to set up a consulting source that would be readily available as a job shop. On the other hand, the primary purpose of MRC was to create a nucleus of applied mathematicians, who conducted research on topics of general relevance to the Army, and who by their presence were a great source of advice to Army scientists and engineers through interactions, collaborations, and education. The role of AMSC, especially in relation to these conflicting expectations, is another topic that we explore in the rest of this monograph. Research and academic institutions today still face the challenges of generating and transferring knowledge and technology that the center faced from its beginning.

Dr. Ivan R. Hershner, Jr.

Ivan Ray Hershner was born in Lincoln, Nebraska, in 1917. His early education was at the University of Nebraska, where he obtained his bachelor's and master's degrees in mathematics. He proceeded to Harvard University in 1940 for graduate work, but the conditions of World War II made it uncertain whether he could complete his doctoral degree. He had been attracted to an article published in *Life* magazine in 1941 on paratroops and had decided then that if he were called to the military, he would serve as a paratrooper. In January 1942, Hershner was called to active duty, and he volunteered for parachute duties. He graduated from parachute school in June 1942 and subsequently served with the 101st Airborne Division at the siege of Bastogne.

In November 1945, Hershner returned to Harvard and began studies toward a Ph.D. in mathematics under Professor Joseph Walsh. In the following spring, Professor Saunders MacLane offered him a part-time teaching position concurrent with his graduate studies. In 1947, he completed a Ph.D. and started his academic career at the University of Chicago. Later, he moved to the University of North Carolina at Chapel Hill (UNC-CH). It was at UNC-CH that Hershner developed his contacts with the U.S. Army Ballistic Research Laboratory (BRL). He spent summers at BRL, along with one of his students. At one point, he had a choice between conducting research in mathematics and learning about computing. Hershner opted for the latter and continued his association with BRL for more than five years. He then went to Raytheon to work on the Patriot missile system, which later played a large role in the Gulf War. During the Korean War, Hershner had been recalled to two more years of active duty as a research and development officer. He continued his association with the Army from that point, first as a consultant and later as a civilian executive until his retirement in 1980.

Early in 1955, the Army leadership wanted to establish an interdisciplinary mathematics group that would conduct basic research in areas of applicable mathematics while simultaneously remaining available to the Army for advice. Hershner, who at the time was chairman of the Department of Mathematics at the University of Vermont, was brought in to lead the effort of generating suitable proposals from the leading U.S. academic institutions, conducting a review process, and making recommendations for an appropriate site. When the University of Wisconsin was selected for the establishment of the Mathematics Research Center (MRC) in 1956, Hershner became the key action officer for the Army. He served initially as the deputy chair of the Army Mathematics Advisory Group (later expanded to the Army Mathematics Steering Committee, AMSC), and later as its chair. Hershner chaired the AMSC until 1973. This committee was the principal advisory group to MRC and provided the key liaison between MRC and Army scientists and engineers.

Conclusion

The history of warfare offers many examples where nonmilitary specialists have provided significant new ideas and tools about the machines, the tactics, and the strategies of war. In particular, it has become clear since World War II that scientific research is absolutely essential to national security. Equally important is the fact that many of the concepts and scientific tools developed in the context of wartime applications can be just as useful to civilian needs. Real success in scientific fields comes, however, from the free play of intellect and from the freedom to work on problems of one's own choice, in a manner dictated by one's curiosity to explore the unknown. Satisfying this important precept was a key assumption in the establishment of MRC.

MRC provided a balanced view of mathematics, in contrast to other trends of the 1950s and the 1960s that emphasized the separation of "pure" and "applied" mathematics. This center was well disposed towards the development of a genuine applied mathematical program that was not only stimulated by practical problems but also encouraged by intellectual curiosity, and that was not constrained by the limitations of any specific applications. Outsiders often consider mathematics to be a tool or a framework. Rarely do people expect new technical developments to arise from work in mathematics. Centers like MRC have proven such attitudes to be, at best, incomplete. Ample evidence for this also comes from current developments in computer science (and their antecedent activities) and from the synthesis of new materials or analysis of structural patterns in biological systems.

Another significant impact of the establishment of MRC was the creation of an excellent environment for students and young researchers. This intellectually rich milieu included visitors from other U.S. and foreign institutions. The greatest legacy of MRC is that it and the Courant Institute of Mathematical Sciences led the mathematical sciences community in the development of a research agenda and in defining directions for research in applied mathematics. At a crucial time in history, when there was a serious shortage of mathematically trained people in the workforce, MRC became an important source for training young scientists in the mathematical sciences.

Chapter 2: The Rosser Years, 1963–1973

Anyone who would now seriously oppose scientific research and development should reflect that consistency in this position would have required him, had he lived at an earlier age, to have opposed the invention and introduction of the use of fire and of the wheel, and the domestication of horses and of elephants. All four of these discoveries have had military applications; and the first two of these are still of military importance. Had there been a prophet in man's prehistory who foresaw the evil applications of fire and the wheel, how likely is it that he could have stopped mankind from acquiring and using the knowledge of them?
— Stephen C. Kleene, Dean, College of Letters and Science, University of Wisconsin, November 1969

Introduction

This chapter covers the way MRC operated during the directorship of J. Barkley Rosser, during whose ten-year tenure MRC evolved into a world-renowned center for applicable mathematics. It was a period that saw both rapid change and enormous stress at MRC, including the bombing of the building in which it was located.

We start with an examination of the research program, considering first the issues of formulation and resource allocation and then surveying the program as it was during the late 1960s to give a sense of how the place worked during what one could reasonably consider to be its best period. Next we turn to the crucial impact of the Vietnam War on MRC's operation and its position on the Wisconsin campus. Near the end of the period, for various reasons including financial stress on the U.S. Government, support for mathematics in the United States declined. We describe a few aspects of this decline and their impact on MRC.

We end the chapter with Rosser's involuntary retirement as director, which marked a major turning point in MRC's history. Both the leadership style and the operating practices of MRC changed markedly, to the extent that in very significant ways it was a different place.

Professor J. Barkley Rosser

J. Barkley Rosser was born in Jacksonville, Florida, on December 6, 1907, the son of Harwood and Ethel (Merryday) Rosser. He received the B.S. and M.S. degrees from the University of Florida at Gainesville, and the Ph.D. in mathematics from Princeton University in 1934. He married Annetta L. Hamilton in 1935, and they had two children, Edwenna (Rosser) Werner and Barkley Rosser, Jr.

After postdoctoral work at Princeton and Harvard, Rosser began his regular academic career at Cornell University, where he was appointed instructor in mathematics in 1936. He rose quickly, attaining the rank of professor of mathematics in 1943. He remained at Cornell in that post until 1963, serving as chairman of the Department of Mathematics from 1961 to 1962. During World War II, from 1944 to 1946 he was chief of the Theoretical Ballistics Division of the Allegany Ballistics Laboratory. In 1963 he was appointed professor of mathematics at the University of Wisconsin in Madison, with a concurrent appointment as director of the Mathematics Research Center (MRC). During his career he held visiting appointments at institutions including Harvard, Princeton, Oxford, the University of Paris, Rockefeller University, and Brunel University. He also served from 1949 to 1950 as director of research of the Institute for Numerical Analysis, National Bureau of Standards, located at UCLA (and was a member of its evaluation committee in 1953), and from 1958 to 1961 as director of the Communications Research Division of the Institute for Defense Analyses (IDA). He retired from the University of Wisconsin-Madison in 1978, with the rank of professor emeritus of mathematics and computer sciences, and died in Madison in 1989 at the age of 81.

Rosser gained distinction as an authority on symbolic logic, rocket ballistics, and numerical analysis; he was the author or coauthor of six books and more than 70 scientific papers. He received the honorary degree of doctor of science from the University of Florida in 1970, and from Otterbein College in 1971. He was awarded Guggenheim and Fulbright fellowships and was elected to membership in the American Academy of Arts and Sciences.

He advised the U.S. Government throughout his career, in areas including rocket ballistics and numerical analysis. He was chairman of the Mathematics Division of the National Research Council from 1960 to 1962, and chairman of the Conference Board of the Mathematical Sciences from 1963 to 1965. His service to the United States was recognized when he was awarded the Presidential Certificate of Merit in 1948, the Navy Certificate of Commendation in 1960, and the Army's Decoration for Distinguished Civilian Service in 1974.

Rosser's numerous professional contributions included serving as president of the Association for Symbolic Logic and of the Society for Industrial and Applied Mathematics, as well as service on numerous advisory groups.

He and his wife Annetta were enthusiastic participants in and patrons of the cultural life of Madison. They participated in the Madison Symphony and its associated Chorus and in the choir of Christ Presbyterian Church. They also supported organizations in Madison devoted to music, art, and protection of the environment.

Shaping the research program

If one is to channel substantial financial resources and the efforts of many people into a coordinated research program, a central consideration has to be how to shape that program. Two key parts of that process are, first, how to determine general areas of work and, second, how to balance that work between what one could think of as consumption and investment. On one hand, the program must produce results directly applicable to the sponsor's concerns, and on the other the program must open up new research fields or further develop existing fields, which may in the future produce new applicable results or new tools for producing such results. Here we comment briefly on the way in which MRC handled these issues during the 1963–1973 period.

Selecting research areas

One might reasonably ask whether MRC used some particular method or set of criteria to make decisions about research areas. Actually, the process was much more subtle than that, at least during the late 1960s and early 1970s. The impetus for support of research subareas, or of particular visitors, generally came from the core of permanent members and others who had long-term associations with the center, such as some of the part-time University of Wisconsin-Madison faculty. These usually started in the form of proposals or suggestions for planning the future research program. Rosser, as director of MRC, had a very strong influence on which of these proposals were to be approved, but in general he did not operate by fiat. Rather, he used his control of MRC resources to channel more funds to proposed areas that he thought were in the best interests of the center's long-term (and sometimes shorter term) needs. Occasionally he might initiate a program himself. The research program thus tended to evolve as the result of a dialogue between permanent members and the director.

In this way the permanent members' energy and interests, and the extent to which they could connect to professional networks that made it possible for them to attract leading workers to MRC, played a key role in the evolution of the program. Those areas having active and influential permanent members tended to have more active research programs and to attract more resources. This method of operation was considerably altered in later years, when the directors of MRC had different personal characteristics and the Army requirements for continuing the MRC contract included the preparation of more detailed proposals than had been the case during the period covered in this chapter. As we will see in Chapter 3, the role of peer review and the evolution of areas of concentration took on increasing importance in the later years of the center.

Broadly based versus focused research

A recurring issue throughout MRC's life was allocation of resources between two competing areas. On one hand was focused research, either on problems of direct interest to the Army or in research areas that could be expected to have direct benefit to Army operations. On the other was broadly based research in applied mathematics, often without specific goals defined in advance, but always in areas judged to be scientifically promising and ripe for development at the time.

Generally, research cannot be neatly classified in one or the other category, either as focused or as broadly based. An example is spline functions, mentioned above. This was a basic research program in the mathematical area known as approximation theory, but the work also produced methods of great usefulness to the Army as well as to many industrial interests. Indeed, when MRC convened the Advanced Seminar on Theory and Applications of Spline Functions in October 1968, representatives of 25 different Army installations attended. When one of the Army installations authorized only one official participant, another employee took annual leave in order to be able to attend.[5]

In contrast to the spline research, which clearly benefited both fundamental research goals and the immediate interests of the Army, some other areas that the center supported fell within the broad contractually defined areas but had much more tenuous connections to problems of interest to the Army. An example was the work in mathematical methods of economics, discussed below. If such areas were of interest to the MRC staff that participated in planning the program, if they seemed meritorious from the research standpoint, and if resources were available, MRC felt free to pursue them.

Connected to the question of focused research is that of direct assistance to the sponsoring agency. Rosser had definite views on this question, and upon his arrival as director in 1963 he changed the method of operation previously used by Langer. The latter had asked certain members to be especially attentive to Army needs and to spend significant amounts of time in helping the Army. Rosser, on the other hand, thought that *every* member should be ready to help with Army problems if needed. He outlined this position very clearly in his May 10, 1963 letter to Ivan Hershner, then chair of the AMSC, at the beginning of his directorship.

> Dear Ray: It occurred to me after our phone conversation the other day that certain of my remarks were made without any explanation of how they fitted into my overall thinking. As a result, you may have quite a wrong impression of my intentions. For that reason, I am writing to supply you with a fuller picture of what my intentions are.
>
> It is essential for the Mathematics Research Center to attract and hold the most capable mathematicians, and to see their talents and experience are most effectively used to assist the Army in coping with its mathematical problems and in upgrading various of its mathematical activities. A key difficulty is the snobbish attitude which many mathematicians display toward applied mathematics generally, and even more particularly toward applications of mathematics in a military direction. Some claim that no first class mathematicians will participate in such an effort. Others say that the best one can hope for is to provide extravagant salaries and particularly pleasant working conditions in return for a limited and grudging help with Army problems.
>
> Though these statements may be true as applied to some mathematicians, they are certainly not true in general. My experience with I.D.A. group at Princeton showed quite conclusively the invalidity of both claims. However, because such claims are freely made, one must make special efforts to succeed in the face of them.
>
> The first thing is to avoid allowing oneself to be infected with such beliefs. Further, one should avoid hiring anyone whose mental horizons are so limited that he gives serious credence to these claims. There is a strong temptation to try to lure some mathematicians of high reputation who do not accept the aims of M.R.C. to join it by compromising and offering them special exemptions from giving proper assistance to the Army. Shortly after I agreed to accept the Directorship of M.R.C., it was very seriously suggested to me by certain people that indeed this is what I should do; specifically, that I

should attempt to hire a few big names by promising them complete freedom from concerning themselves with the Army in any way, and that I should then try to fulfill the obligations of M.R.C. toward the Army with younger and transient staff. Since it would be clear to these younger men that they were considered second-class citizens, it would become increasingly difficult to find competent staff for this purpose, and M.R.C. would more and more fail in its mission. Accordingly, I rejected this suggestion, and have been trying to make it clear to all concerned that I shall expect all employees of M.R.C., from the most distinguished to the neophytes, and whether temporary or permanent, to be concerned with helping the Army with its problems, and to consider this an honorable and worthy concern.

I have two reasons for expecting to succeed with such a policy. One reason is my conviction that an applied mathematician benefits from contact with practical problems; those who do not accept this principle are not genuine applied mathematicians and M.R.C. would be better off not to employ them. The second reason is that the I.D.A. operation at Princeton was set up according to this policy, and has been quite successful. In order to have this policy accepted by the staff, it will not suffice merely to announce it loudly and frequently. It must indeed be a fact that helping the Army with its problems is a worthy activity. If first class mathematicians find themselves challenged by the Army problems, which they encounter, no propaganda will be needed to convince these mathematicians of the worth of those problems. They will have the evidence with which to refute those who claim it is demeaning to concern oneself with the Army's problems. I feel I can substantiate this point by appealing to my I.D.A. experience.

However, one must avoid a pitfall in this direction, namely of succumbing to pressure to take over more and more routine problems from the Army. The Army is a large and multifarious operation, and needs the solution of problems in a quantity that would swamp a facility a hundred times the size of M.R.C. Also most of these problems could be solved in house, by mathematicians closely acquainted with the various non-mathematical matters related to these problems. Otherwise, there is a danger of irrelevant solutions being promulgated, or at best various delays and inefficiencies. The M.R.C. must expect to be called on for help with problems requiring special mathematical techniques and to elucidate these techniques or training. Even then, M.R.C. should be called upon not so much to solve the problems in detail as to discover or select the needed techniques and to elucidate these techniques to those particularly concerned who can then use them to proceed with the solution of their problems.

Similar remarks apply to the use of M.R.C personnel to deliver lectures. Insofar as the material for the lecture or its organization is novel or unusually difficult, and requires the services of an especially experienced or talented mathematician, the members of M.R.C can feel that there is worthy challenge, and can undertake the preparation and delivery of these lectures with pride and a sense of accomplishment. If the lectures are particularly helpful, a reasonable number of repetitions of them will be willingly undertaken. However, it is not an effective use of the talents of the research mathematicians hired at M.R.C. to call on them for wholesale delivery of standardized lectures. If this is desired, additional members should be hired for this purpose, and they should be selected on the basis of exceptional skill in teaching rather than exceptional talent in research.

At the two meetings, one of A.M.S.C. and one of its panel, there seemed a tendency on the part of some members to press for a wholesale teaching function for M.R.C. I repeat what I said above, that some teaching activity must be expected from members of M.R.C., and a reasonable amount should not be objected to. I would prefer not to make an issue of this point in the A.M.S.C. unless they should press a demand for an unreasonable amount. The reason for my mentioning the matter to you was a hope that with some discreet guidance from you the affair could be kept within reason without making an issue of it.

Sincerely, J. Barkley Rosser

Rosser's policy was to be followed by the succeeding directors, right up to the end of MRC's existence 23 years later. Each letter of appointment contained the sentence, "Your advice may also be sought on mathematical questions within your areas of interest and expertise, but not to a degree that this would become burdensome." Although people may certainly hold a variety of opinions about the best way to organize help to a sponsoring agency, one can certainly say that Rosser's method worked well enough to satisfy both academics and the Army over a fairly long period of time.

A view of the research program at work

The period from the mid-1960s to the mid-1970s saw a substantial development and maturation of the MRC research program. In this section we try to give a picture of that program as it operated during the years 1967–1973. We do not give similar reviews for the research program in other periods of MRC's existence, not because those periods were not important but because looking at one period, in what might reasonably be regarded as a high point of MRC's functioning, should suffice to give the reader a picture of how the place worked. In the process we consider some of the staffing issues that appeared during this period. Those issues differed in degree, but not in kind, from issues MRC had already faced in its formative phase.

As in previous years, the basic contract governing MRC specified four principal areas in which research was to be carried on: numerical analysis, applied mathematics, statistics and the theory of probability, and operations research, though MRC's work was not to be limited to these. We first examine how the program areas evolved during this era.

Numerical analysis

Work in numerical analysis had several foci. One of the main areas of work in numerical analysis had probably never been envisioned at the time of its founding. This was the development of *spline functions,* piecewise polynomial functions that can be used to interpolate data, providing a curve that appears smooth to the eye, with only gentle changes in curvature, while passing exactly through prescribed points. Traditional interpolation methods had generally used polynomials, which lack the curvature property. By contrast, a spline consists of different polynomials, pieced together carefully in such a way as to retain much of the simplicity and mathematical convenience of polynomials while providing desired smoothness and good appearance. We will return to this topic more thoroughly in Chapter 4. It suffices to note here that spline functions now play very important roles in a great many areas, ranging from industrial design through computer graphics to statistical data fitting. However, until the mid-1960s they had been regarded somewhat as curiosities, and their theory had not been systematically developed.

Work on splines at MRC started in 1965 under the leadership of two permanent members, I. J. Schoenberg and T. N. E. Greville. The work evolved into a separate area in 1966 and continued for years thereafter. In fact, it probably is the case that spline functions are one of the best recognized of the mathematical advances that MRC brought about. By 1968 significant applications had already appeared, one example of these being the doctoral dissertation of F. R. Loscalzo, comprising Technical Summary Reports (TSRs) 723, 842, and 869. Loscalzo developed a new kind of predictor-corrector method for solving ordinary differential equations, using spline functions. This method turned out to give good computational results for some so-called "stiff" differential equations, which are difficult to solve by other methods.[6] In addition to Loscalzo's work, numerous other TSRs appeared in 1968, including several by Schoenberg and others by L. L. Schumaker. The pace accelerated in 1969, with Schoenberg and Greville continuing their work joined by visitors Samuel Karlin and Zvi Ziegler.

Although the work in spline functions had originally been planned as a three-year "special study,"[7] by the end of this period it had had such obvious success that the center continued it. Splines were therefore perhaps the first example of an "area of concentration," deliberately planned to generate intense activity in a subfield of research for a predetermined time period. As we see below, in later years these areas of concentration were to become a formal part of the MRC planning process (see Chapter 3), partly in response to reductions in resources.

In addition to advancing research in splines, MRC also took care to see that the results were disseminated. The area was judged so important that MRC devoted two semiannual conferences to it: the Autumn 1968 Advanced Seminar on Theory and Applications of Spline Functions, chaired by T. N. E. Greville, and the Spring 1969 Symposium on Approximation Theory with Particular Emphasis on Spline Functions, chaired by I. J. Schoenberg.

Activity in the numerical solution of ordinary and partial differential equations and of integral equations continued during most of this period, but in the early part of the period the level of effort was not uniformly high. In 1968, eighteen TSRs were written on these topics,[8] but in 1969 the area was lumped together in the MRC Annual Report with other efforts in numerical analysis such as work on ill-conditioned matrices and computation of zeros of the Riemann zeta function. Numerical analysis activity picked up somewhat in 1971–1972, with visits by James Bramble (Cornell) and Vidar Thomée (Chalmers Institute of Technology, Sweden). In 1973 MRC made strong efforts to increase the level of activity, with visits for parts of 1973–1974 by Jim Douglas, Jr., Heinz-Otto Kreiss, J. A. Nitsche (Freiburg im Breisgau, Germany), and Mary F. Wheeler. The rather disjointed pattern of activity seen here was fairly typical of MRC's operation in numerical analysis; with the notable exception of spline functions there was no sustained, coherent effort leading to substantial accomplishment over a period of years.

One reason that may account for this was the fact that MRC never succeeded in recruiting a permanent member to lead and coordinate efforts in this area (again, with the exception of splines). Visitors came and went, and they were often people of very high quality. However, there was no permanent member to provide continuity and persistence in the program. Rosser certainly was a numerical analyst of note, but his energies during this period were almost completely absorbed by the duties of the directorship. Ben Noble and Louis B. Rall also made effective contributions to topics in numerical analysis, as did T. N. E. Greville; however, again, other areas absorbed much of their attention. The contrast between the sustained success of the spline function subarea (benefiting from the continuous attention and organizational work of Schoenberg and later of Carl de Boor) and the sporadic nature of the other numerical analysis work provides a striking example of the importance of influential continuing staff in the development and sustenance of a research area.

Another important aspect of the program in numerical analysis at the center was the impact of functional analysis. Applied functional analysis (perhaps more in the spirit of the Kantorovich school) had a significant impact on the development of a rigorous framework in which numerical schemes were developed, analyzed, and critically evaluated. We will return to this topic in a more elaborate way in Chapter 4.

Statistics and probability

Activity in probability and statistics showed variable intensity over this period. In contrast with numerical analysis, there was a permanent member, Bernard Harris, in place during the entire period to oversee and guide the area. One of Harris's main interests was reliability, and another was the use of combinatorial methods in statistics. The pattern of activity in probability and statistics at MRC reflects both of these interests.

In 1968, twelve TSRs were written in various areas of probability and statistics, including reliability, probability distributions on algebraic structures, time series, Bayesian inference, stochastic processes, and information theory.[9] This activity did not include a notable accomplishment of that year, namely, the solution by M. Fox and G. S. Kimeldorf of the problem of noisy duels. We describe that below, in the section on operations research.

In 1969, the center's annual report commented, "No one central activity was organized in this area, but a number of small groups pursued somewhat diverse topics." It went on to report the production of nineteen TSRs in subareas including reliability, stochastic integration, stochastic processes including branching processes, multivariate analysis, nonparametric methods, and others.[10] In the following year (1970) the director noted in his report on future plans that "probability and statistics will be weaker than desired, because of several disappointments in recruiting, as well as the impending loss of Professor H. B. Mann."[11] He went on to report, however, that MRC was making efforts to attract an eminent authority on reliability, Frank Proschan (then at Boeing Scientific Research Laboratories but on leave to Stanford University) as a permanent member. He noted that he could make no estimate of the prospects of success, and in fact the effort turned out to be unsuccessful. The area continued to have a relatively low profile in MRC during the rest of the period covered by this chapter, with steady work by Harris and augmentation by visitors. The latter included various members of the university's Statistics Department, who accepted part-time research appointments that MRC offered in order to increase the continuity of the research work in this area. It is worthwhile to note that from the point of view of the user community (in this case the Army), this area continued to be a major topic of interest.

Mathematical analysis and physical mathematics

The original contract negotiations envisioned that the area of mathematical analysis and physical mathematics would include ordinary and partial differential equations (theory, as distinct from numerical solution methods), fluid mechanics, elasticity, plasticity, electrodynamics, electrical networks, wave guidance and propagation, and other topics. It was thus a large and significant part of MRC's work, and one that intersected many domains of interest to the Army. The founding director of the center, as well as many senior members and visitors at the center, had strong backgrounds in one or more of the subfields in this area.

In the late 1960s, the work in progress in this area included studies in fluid dynamics and non-Newtonian liquids, applied partial differential equations, variational methods, dual series methods for partial differential equations, and buckling of shells. As an indicator of the level of activity, MRC produced twenty TSRs in this area in 1968 and fourteen in 1969. During this period Ben Noble, a permanent member of MRC who oversaw and guided much of the activity in this area, was in residence. However, during the academic year 1969–1970, Noble had begun to disengage somewhat from the MRC applied mathematics program to concentrate on other interests, including writing plans. For instance, in the period from October 1969 through September 1970, his name appeared on no TSR.[12] In the spring semester of 1971, he went on leave to Oberlin College to participate in an educational videotape project supported by the National Science Foundation, involving the videotaping of lectures from one of his books.[13] In the fall of 1971, he went on a two-year leave to Oxford University with plans to write a monograph on integral equations. He returned for a semester in the fall of 1973 and then left again for the first half of 1974.[14]

Although Noble was not physically present during much of the latter part of the period covered by this chapter, work in applied mathematics nevertheless continued at a fairly strong level due to the part-time appointments of several University of Wisconsin-Madison faculty, including Arthur S. Lodge (rheology) and Warren E. Stewart (fluid mechanics and reactor design) of the College of Engineering, as well as the efforts of a number of visitors.

Operations research and related areas

Operations research and areas related to it had three permanent members during at least part of the period covered by this chapter: T. C. Hu, H. F. Karreman, and J. B. Rosen. Hu was active in combinatorial optimization and network flows, and Rosen in nonlinear programming, but it is difficult to characterize the research emphasis of Karreman: during the entire period of his appointment at MRC his name appears on only one MRC TSR (No. 653, coauthored with Harris and Rosser in 1966). There were also faculty from the University of Wisconsin-Madison on part-time research appointments, including O. L. Mangasarian, R. R. Meyer, and S. M. Robinson; the latter also served as assistant director of MRC from mid-1971 through 1974.

Visitors did much of the work produced in this area from 1967 to 1973. As mentioned earlier, a notable example early in this period was the solution by M. Fox and G. S. Kimeldorf of the problem of noisy duels. This was an example of a kind of problem that was interesting to both mathematicians and military tacticians. It was described as follows in MRC's annual report for fiscal year 1968:

> Roughly speaking, the problem is as follows. Two opponents start with a pistol apiece and a number of rounds of ammunition (not necessarily the same for the two opponents). The[y] approach each other, firing at will. Each has a probability p(x) of hitting the other if he fires at a distance x (not necessarily the same probability for each opponent). The duel is called "noisy" because each hears if the other shoots (a "quiet" duel could be submarines firing torpedoes at each other, where one might not know if the other fired and missed). Given the number of rounds that each opponent starts with, their initial distance apart, and their respective probabilities of hitting, what is the best strategy for each opponent and what is the probability of survival of each opponent under this strategy?
>
> To study such a game mathematically, Fox and Kimeldorf had to make a number of restrictions. The man who runs out of bullets first isn't allowed to duck out. Also, once a gunman is hit, he's dead.
>
> Then if the less accurate duelist is given many more bullets than his opponent, he can shoot first and let fly with so much lead that by sheer chance he can make a hit. Just how many bullets are needed is enormous, the two mathematicians found.
>
> In the research, Fox and Kimeldorf computed a number of duels with many variations so as to tabulate the price of inaccuracy.

One can roughly classify work in operations research into two main areas: optimization and applied probability. Within these, the optimization area contains subareas including continuous optimization, comprising linear and nonlinear programming, and combinatorial (discrete) optimization, dealing with problems in which the variables have to take discrete (possibly integral) values. An example of the latter is the assignment of aircraft to routes; one cannot fly half an aircraft, so the assignments must be in integral numbers. The continuous optimization area also encompasses some other distinct fields, such as (deterministic) game theory and the treatment of certain economic models. Similarly, the field of applied probability contains further subareas, such as queuing models, simulation, and inventory problems.

The work of Fox and Kimeldorf was an example of applied probability combined with game theory, and in that way it was notable not only for the solution of a difficult problem but because it spanned different areas. This was not the case with most of the operations research work that MRC carried out, which fell within one or another of the recognized subdivisions.

Among these, applied probability was actually very little represented, though it is an important field. Perhaps one reason for this was the distribution of interests among the active permanent members. Hu specialized in combinatorial optimization and Rosen in deterministic nonlinear programming. The part-time staff from the University of Wisconsin-Madison (Mangasarian, Meyer, and Robinson) all worked mainly in deterministic optimization. Not surprisingly, the bulk of MRC's work in operations research consisted of investigations in combinatorial and in nonlinear optimization. Distinguished investigators, including Selmer Johnson of RAND and Garth McCormick of the Research Analysis Corporation, visited MRC to participate in this work. Postdoctoral research workers, such as Lynn McLinden and Melanie L. Lenard, joined them.

A partial exception to this was a sustained effort, organized in part by S. M. Robinson after he joined the research staff in 1970, to broaden the work to include solution of problems related to economic models. These problems tended to use tools from nonlinear optimization, but they required application of those tools in ways somewhat different from those familiar to investigators in nonlinear programming. They also led to connections with research staff from varied backgrounds and somewhat different interests than those of most operations research workers. Visitors of this type, some coming from economics departments or economic research institutes, included Jean-Pierre Aubin, Richard H. Day (University of Wisconsin-Madison Department of Economics), B. E. Easton, B. Stigum, Hiroshi Konno, Jerzy Łoś, and Maria Wycech Łoś. This effort later became a casualty of the resource shortage and the consequent refocusing of work in areas of more evident interest to the Army.

Impact of the Vietnam War on MRC

Few institutions in American life were unaffected by the war in Vietnam, and MRC surely was not among those. The war brought nothing but trouble, first in the form of disaffection on the part of various members of the faculty who had contributed to MRC's development, then in growing agitation and opposition by other faculty, students, and hangers-on in the campus community, and finally in a bombing attack against the building that housed the center. The "student troubles" that plagued the campus from 1968 to 1971 were due to opposition to that war, which was widespread among students and others in Madison. The "MRC troubles" were, however, caused primarily by a small number of activists.

In this section, we examine first the development of opposition to MRC, which in some cases served as an outlet for the anger and frustration of those who opposed the war but could do little about it. We start with the picture painted by the anti-MRC militants of MRC's aid to the Army and then contrast it with the actual work that MRC did. We then describe the ill effects on the center of this sustained and vigorous campaign of opposition. These effects intensified over time until August 1970, when the center's building was bombed.

MRC as a focal point for opposition

At the end of the 1960s, the Madison campus contained many who were disaffected from one or another part of the contemporary scene. Some had relatively specific disagreements with the university or with other perceived repositories of authority. Others did not endorse the directions being taken by American society, by the U.S. Government, or by other entities operating on a much broader scale than did the university. It is important to remember that this was not something unique to Madison, nor did it stem entirely from the Vietnam War. Very similar things were simultaneously happening in countries that were not then entangled in Vietnam, such as France, Germany, and Italy. However, in Madison the most visible aspect of this dissatisfaction surely was the well-organized, sustained, and vigorous opposition to the Vietnam War and to every agency perceived as helping the prosecution of that war.

Among such agencies, MRC occupied a special place. It was, in a certain sense, a foreign body in the university. It was not an academic department, since it did not have any students except for graduate students enrolled in other departments who happened to be studying under the guidance of MRC staff. It also drew its sustenance from a contract with what was perceived to be the villain of the Vietnam scenario, the Department of Defense. Therefore it was a natural focus for agitation on the part of disaffected students and of those hangers-on about the campus community who wished to associate with students or to use students for their own purposes. However, during this period, such agitation was not confined to MRC. For example, the Land Tenure Center (another contract research center on the Madison campus) also experienced threats and disruptions.

If one opposes an organization and wishes to convince others to share one's point of view, a way to accomplish that aim is to portray the organization as doing things that one's audience finds abhorrent. Therefore, the arguments marshaled by those opposing MRC tended to focus strongly on the specific connections of MRC to the Army in order to persuade others that MRC was somehow directly responsible for enabling those actions of the armed forces that members of the campus community could be expected to dislike.

Among these connections, the contractually mandated assistance that MRC provided to the Army with the solution of mathematical problems encountered by Army agencies must have seemed especially tempting. If this connection could be exploited to convince listeners that MRC somehow was doing things that did not belong on a campus, then those making the arguments might attract more support for MRC's removal. Another equally provocative issue was the perceived requirement for security clearance for permanent members. This was a suitable issue with which to stir emotions, since it was argued that if clearance was necessary, then members must have been working on secret projects as part of their university duties. There was also the implication that the university was taking outside direction about who could be supported with MRC resources.

During the years of protest against the war and against MRC, many persons wrote documents, pro or con, about the center's activities in support of the Army. Among all of these, the one that stands out as probably the most comprehensive single presentation of the case against the center is a booklet called *The AMRC Papers,* produced in 1973 by a group calling itself the Madison Collective of Science for the People.[15] The foreword to this document portrayed the center (to which it consistently, and incorrectly, referred as AMRC) as engaged in "a struggle between the peoples of the world fighting for liberation and the death machine of the United States."

This document runs to some 119 pages, not counting the front matter, and we certainly cannot summarize it here. However, we can provide an idea of its organization, the arguments that it makes, and what appears to have been the mental orientation of those who prepared it. This should suffice to give an idea of what the anti-MRC faction was trying to communicate.

The booklet is organized in four parts, whose titles are

- How AMRC Helps the Army
- How AMRC Works
- AMRC's Relationship with the University of Wisconsin
- An Alternative: A People's Math Research Center

The part of most interest here is the first, which includes four chapters on specific areas in which it is alleged that MRC helped the Army. The titles of these chapters are Counterinsurgency, Chemical & Biological Warfare, Missiles, and Conventional Weapons. In reading through these, one finds that they are for the most part straightforward reports of consultation, based wherever possible on documentary sources that are carefully cited. Indeed, many of the descriptions reported in these four chapters are taken directly from the reports of the center itself, and others from documents produced by military agencies. As an aside, it is interesting to note that this booklet was being sold in Madison at a time when some Army scientists responsible for oversight of the MRC contract were in town. Mindful of the difficulty they frequently encountered in persuading other Army officials that mathematical research was doing anything of real use to the Army, the scientists went out on the street and bought 40 copies of the booklet because it made such powerful arguments that MRC was in fact of great benefit in advancing the Army's programs![16]

Why did the authors of this booklet believe that a factual presentation of MRC's work in support of the Army would help their cause? One answer is that they seem to have had a mindset in which anything that helped the Army necessarily harmed "the people," whoever those might be. The following statement illustrates this mindset:

> [A]s long as this research is directed toward the needs of the military instead of the needs of the people, it cannot be said that AMRC is serving the public. AMRC's newest research for social and economic manipulation can only be stopped by political action from people opposed [to] the imperialists' use of science.[17]

To view this piece as an argument against MRC seems to require a certain predetermined political viewpoint. Madison at the beginning of the 1970s contained many people with just such a viewpoint, and some of them were ready to use violence to attain their ends. In the early hours of August 24, 1970, terrorists exploded a car bomb next to Sterling Hall, the university building in which MRC was housed. The explosion killed a physics researcher who was working on an experiment (not in MRC's part of the building) and caused a considerable amount of damage to the building. It did relatively little damage to the books, papers, and office equipment contained in MRC. Indeed, within a very short time the MRC operation had been relocated to temporary space in several locations on the campus, and work was again under way.

However, although the bombing itself had little substantive effect on MRC's research program, the atmosphere on campus and the tensions caused by the antiwar activities and the responses to them (including riots, tear gas, and fire bombings) had a serious impact on the morale of the MRC staff and on the relationship of MRC to the rest of the campus. In the next section we describe some aspects of this effect.

Effects on the MRC program

Effects of the war on the MRC staff had been felt as early as 1968, when Professor Donald Greenspan resigned from MRC and moved full-time into the Computer Sciences Department.[18] As time passed, things became more difficult, both in the matter of attracting and retaining staff and in that of maintaining good relations with other campus departments. The constant pressure of vocal opposition and the tactics of groups opposed to MRC sometimes made it difficult to maintain the kind of environment conducive to good research.

An amusing example of those tactics was the appearance at the entrance to MRC one day of a group of young people led by one of the priests from a local Episcopal church. They carried a coffin, which they set down just outside the door. The priest then performed a religious service, which interested onlookers (many from the MRC staff) were told was a service of exorcism—presumably of the manifold evils that MRC was thought to bring to the campus. A much less amusing example was the appearance outside the same door of a mob that proceeded to throw rocks at the center's offices, breaking windows and subjecting some of the staff to considerable personal danger. After that incident the window glass was replaced with Lexan, but the sense of insecurity remained.

During the tension following the bombing, two of MRC's permanent members left the center, one for extended leave and one permanently. As was mentioned above, Ben Noble went on extended leave beginning in 1971. J. B. Rosen, a permanent member in operations research, left the university in mid-1971 to join the University of Minnesota.[19] In addition to these departures, there were other invisible losses in the form of declined appointments and reduced interest in visiting MRC. The extent of these losses will never be known, but they were certainly important.

Another event of the period immediately after the worst troubles was the Madison campus's assertion of greater control over the internal functioning of MRC. In 1972, the chancellor of the University of Wisconsin-Madison constituted a permanent Advisory Committee to MRC. This committee of eleven members represented a wide range of constituencies across the campus. In addition to MRC there were to be members representing the campus administration, the three faculty divisions of Biological Sciences, Physical Sciences, and Social Studies, the Graduate School (which at Wisconsin is in charge of research policy and administration), the College of Letters and Science, and the departments of Mathematics (two members), Statistics, and Computer Sciences. It replaced an ad hoc committee that had been in operation under the leadership of Professor J. O. Hirschfelder of the Department of Chemistry, and it was intended "to advise the director of MRC on permanent appointments and major policy matters." As no such advice had previously been considered necessary during the 16 years since the signing of the MRC contract in 1956, the formation of this committee constituted a significant event in the relations between MRC and the university administration. For the first time, the director of MRC—who was not to be a member of the advisory committee—was to have a permanent, formal oversight group appointed by the campus in addition to the oversight that the Army had provided through the AMSC.[20]

Declining support for mathematics

Although mathematics has been a vital factor in the development of industry, commerce, and public works in the United States, this period saw a substantial decline in public support for mathematics. We briefly review some reasons for this decline, and we trace the effects on MRC of what was happening in the society at large.

In the late 1960s and early 1970s federal research and development support became increasingly mission-oriented. The 1969 Mansfield Amendment accelerated this trend, causing a considerable reduction in Department of Defense support for mathematical research. Along with these changes came very damaging changes in graduate student support.

In the words of the National Research Council's (NRC) Ad Hoc Committee on Resources for the Mathematical Sciences:

> Dramatic reductions in federal fellowships beginning in 1971 removed virtually all federal support of mathematics graduate students and postdoctorals. Compensation for these two types of losses could only be made at NSF, but at NSF constant dollar support of mathematical research decreased steadily after 1967. We estimate the loss in federal mathematical funding to have been over 33% in constant dollars in the period 1968-73 alone; it was followed by nearly a decade of zero real growth, so that by FY 1982 federal support for mathematical sciences research stood at less than two-thirds its FY 1968 level in constant dollars.[21]

This sharp decline in support for mathematics was not immediately recognized for what it was. The NRC's Ad Hoc Committee identified three reasons that it thought had contributed to this delayed recognition:

- "After the sharp decline of 1968-73, universities increased their own support for many things which earlier would have been carried by research grants. It was only after financial problems hit the universities in the early 1970s that the severe lack of resources became evident.
- "The growth in computer science support masked the decline in mathematics support because of the federal budget practice of carrying 'mathematics and computer science' as a line item until 1976.
- "The explosion of the uses of mathematics caused funding to flow into applications of known mathematical methods to other fields. These were often labeled "mathematical research" in federal support data. The category grew rapidly, masking the fact that support for fundamental research in the mathematical sciences shrank."[22]

MRC and the Madison campus, like other institutions, did not at first realize the extent of the change they were experiencing. They did, however, see that funds were tighter than they had previously been. Funding limitations made it necessary to reduce the number of new Ph.D. graduates brought to MRC for postdoctoral work.[23] These postdoctoral researchers had always been an important part of the program, both because they contributed intellectual energy and the latest research training, and because most would later go to other universities and take with them the knowledge and viewpoints about applied mathematical research that they had gained at MRC. Therefore, the MRC staff thought it of considerable importance to try to restore this part of the program.

For this reason, MRC prepared a proposal to the NSF for support of several postdoctoral research workers who would visit the center for two-year terms.[24] This was the first time that MRC had gone outside the framework of the Army contract to support its research staff, and it marks both a shift away from exclusive reliance on the Army and a first step toward dealing with the funding limitations that were to become much more severe in the years that followed. The NSF proposal was in fact to be approved, after some delay, and it provided much-needed support for newly trained research workers to contribute both to the MRC program and to their own professional development.

Change in the MRC leadership

By the early 1970s, in the minds of many both on the Madison campus and around the world, J. Barkley Rosser symbolized MRC. He had led the center since 1963, and in that time he had brought many outstanding mathematicians from all over the world to join the MRC program for varying periods of time. Therefore, he was both well known and highly respected by many influential leaders in applied mathematics. His personal style was refined and polite (to an extent described by more than one visitor as "courtly"), but at the same time was vigorous and was not marked by shyness. He directed the center with a firm and steady hand and was closely involved in all major decisions.

In such a situation, the relatively sudden loss of Rosser's leadership caused substantial change both in the trajectory of MRC and in its internal relationships. In this section we first describe the situation surrounding Rosser's retirement and then look at some of the immediate effects that it had on MRC's leadership. We deal in other chapters with other effects further removed in time.

The retirement of J. Barkley Rosser

For many years the University of Wisconsin-Madison had in place a policy that senior administrators must retire from their administrative posts by the end of the fiscal year in which they attained age 65, though they could retain their tenure as professors past that time. Until 1972, as far as anyone connected with MRC knew, this policy had never been applied to anyone below the level of dean. MRC had the administrative status of an academic department within the College of Letters and Science, and the position of its director was therefore somewhat analogous to that of a department chair, though the directorship was not filled in the same manner (that is, by election) as is that of a chair at Madison. Rosser's predecessor, Rudolph Langer, had retired from the MRC directorship at age 69.

Rosser and others at MRC were therefore extremely surprised when he received a letter in 1971 informing him that, as he would attain age 65 in December 1972, he would have to retire from the directorship by the end of June 1973. As this appeared to be the first time that this policy had been applied to someone at his level, some at MRC concluded that the Madison campus administration had found a device to remove an inconvenient symbol of troubles that they were anxious to put behind them.

This suspicion was strengthened by the undeniable fact that Rosser had led MRC to great scientific distinction during his tenure. He had, however, become an object of personal hatred to many in the community because of his unflinching leadership in the face of antiwar agitation and attacks, and it seems plausible that the administration may simply have wanted him out for local political reasons and may have been willing to use any means at hand to accomplish that end.

A search and screen committee under the chairmanship of R. C. Buck of the Department of Mathematics was charged with identifying suitable candidates to succeed Rosser as director.[25] It proceeded with its work, and in May 1972 Rosser wrote to Hershner, then the chairman of AMSC, to report that, "This committee will have the names of potential candidates to submit to AMSC and the Advisory committee to MRC this spring, so that actual negotiations can proceed. Professor Rosser will retire as Director at the end of June, 1973."[26]

In fact, the committee recommended the appointment of V. L. Klee, Jr., of the University of Washington, as Rosser's successor, and in September 1972 the Budget Committee of the Department of Mathematics approved a tenured appointment in that department for Klee, along with leave to serve as director of MRC. However, Klee subsequently declined the offer. With the deadline for Rosser's retirement fast approaching, the chair of the search committee, R. C. Buck, was asked to serve as acting director beginning July 1, 1973. He agreed to do so for a term of two years, or less if a new director could be found before that time.[27]

Change in the style of leadership

The appointment of R. C. Buck to the acting directorship of MRC began a significant change in the role of the director, though neither Buck nor others at MRC knew that at the time. In fact there were to be no more outside directors; from that point until the nonrenewal of the MRC contract, its directors were to be appointed from within the Mathematics Department on the Madison campus. This caused a significant shift in the relationship of the director of MRC to others on the campus.

As we have noted above, an Advisory Committee had already been put in place to strengthen the control of the campus community over major decisions at MRC. Having a series of directors who were all long-serving members of the Mathematics Department could only increase that influence. Rosser had been appointed from the outside, coming to the campus as director of MRC, and had had the opportunity to negotiate his terms for the position with no pre-existing relationships or obligations to impede him. Those who succeeded him were in a very different situation. All had had long relationships with colleagues and with the campus administration, and now they were being asked by their dean to take an administrative position under his authority. All had gone through the troubles on the Madison campus and had long personal acquaintance of colleagues who strongly opposed some or all of the activities, and even the continued existence, of MRC. This was a very different position from that in which Rosser had found himself when he arrived in 1963.

For that reason the point in mid-1973 when J. Barkley Rosser left the directorship marks a change both in the way MRC operated and in its relationships to other units on the Madison campus. In the next chapter, we examine how MRC tried, with some success at first, to adapt to the new realities of the research environment, funding limitations, and campus relationships.

Conclusion

Fortunately for MRC, the decade-long tenure of J. Barkley Rosser provided the right kind of leadership to accomplish the daunting task of shaping the MRC program. During this period, MRC evolved into a world-renowned center for applicable mathematics. While the core or permanent members of the center initially played a key role in the evolution of the program through their energy and interest and through the extent to which they could connect to professional networks to attract leading researchers to the center, this method of operation was considerably altered in later years. As the next chapter will reveal, the role of peer review and the evolution of areas of concentration took on increasing importance in the later years of the center.

A recurring issue throughout MRC's life, which Rosser very deftly handled during the formative and critical years of the center, was the allocation of resources between two competing areas. One of these was focused research, either on problems of direct interest to the Army or in research areas that could be expected to have direct benefit to Army operations. The other was broadly based research in applied mathematics, often without specific goals defined in advance, but always in areas judged to be scientifically promising and ripe for development at the time. Rosser was well aware of "the snobbish attitude which many mathematicians display toward applied mathematics generally, and even more particularly toward applications of mathematics in a military direction," as he had put it in his 1963 letter to Hershner. He should be credited with striking the right balance in developing a forward-looking research program while at the same time developing an excellent rapport with the sponsoring agency. For instance, he convinced the Army that MRC should be called upon, as he put it, "not so much to solve problems in detail as to discover or select the needed techniques and to elucidate these techniques to those particularly concerned, who can then use them to proceed with the solution of their problems."

This period also witnessed the most tumultuous time in the life of the center. Few institutions in American life were unaffected by the war in Vietnam, and MRC surely was not among those. The arguments marshaled by those opposing the existence and operation of MRC on the campus tended to focus on the specific connections of MRC to the Army in order to persuade others that MRC was somehow directly responsible for enabling the Army's actions.

Although mathematics has contributed significantly during both war and peace, and it has been a vital factor in the development of industry, commerce, and civil work, this period in the United States saw a substantial decline in public support for mathematics. This presented major challenges to large multi-investigator groups like MRC.

Another major change in the operation of MRC was the departure of Rosser from the directorship of the center. The appointment of R. C. Buck to the acting directorship of MRC marked a change in the role of the director, the most salient aspect of which was that all future directors would be chosen from within the university. This inevitably and significantly affected MRC's operation as well as its relation to other campus units.

Chapter 3: Transition and Endgame, 1973–1987

The most vitally characteristic fact about mathematics is, in my opinion, its quite peculiar relationship to the natural sciences, or, more generally, to any science which interprets experience on higher than purely descriptive level.
—John von Neumann, in "The Mathematician," *Collected Works,* 1961

Introduction

This chapter covers nearly half of MRC's 30-year life; however, during that period life at MRC had quite a different aspect than it had had in the early years. For much of this period the directors of MRC had lost the administrative initiative. They spent much of their time trying to adjust the MRC program to their conception of what the Army was looking for at the moment, rather than setting their own directions as Langer and Rosser had largely done.

This is not a criticism of the later directors. As we will show, the research environment had changed. For one thing, the ARO was under new pressures from within the Army itself, having become part of a commodity command (now called the Army Materiel Command) rather than being administered directly from Washington as it had been in the past. For another, funds were lacking, partly because of shortages in the aftermath of the agonizing and divisive war in Vietnam, partly also because of significant inflation, itself closely related to the way in which the government had paid for that war. In the aftermath of the war, many in public life were demanding new levels of accountability from government agencies, perhaps partly as a result of the lack of trust engendered by the war. For all of these reasons, ARO exercised much closer supervision over MRC than it had done in earlier times.

This is also not to indicate that research at MRC was in any way of lower quality or quantity than it had been before. In fact, one could make a good case that the work at MRC during the second half of its life, at least in the fields of analysis and approximation theory, was the best that it had ever done. Indeed, in the fall of 1977, Noble, then serving as director, observed in a report to Chandra that:

> By common agreement here, 1977/'78 will go down as one of the great years in MRC history. Professor M. G. Crandall, with the help of Professor John Nohel and many others, has assembled a very strong group of researchers in time-dependent partial differential equations (some would claim the strongest group anywhere in the world at the moment). Our postdoctoral appointees in this and other areas are also outstanding.[28]

Thus, the problems we are discussing here were not with the research effort itself but with the financing and administrative direction of that effort, and moreover they were problems for which no single person or organization was to blame. Conditions for the support of research—nationally, within the Army, and at Madison—had simply changed. Those changes were to make life progressively more difficult for all parties involved, until they culminated in the demise of the center.

In the sections that follow, we sketch some of the main events of this period, with heavy emphasis on the organizational methods that MRC adopted in order to introduce additional flexibility into its program and to adapt that program to changes in the external research environment. We start with the evolution of what MRC came to call "areas of concentration," which were basically devices for steering increasingly scarce resources into particular areas that could change from year to year to provide a measure of responsiveness to current opportunities and to ARO's guidance. This leads naturally to consideration of a totally new element, external peer reviews, which first appeared at MRC during this period. We discuss the first of these reviews and the far-reaching consequences of its results. We examine some of the structural rigidities that MRC encountered in trying to change from its accustomed ways of organizing and conducting research in order to respond to the new external pressures. Finally, we explore the end of MRC, including a second peer review, which gained it another three years' support, and the final contract solicitation. For the first time, the contract was put out for competitive bids, and we describe Wisconsin's unsuccessful attempt to compete against other universities to retain the Army contract.

Professor Ben Noble

Ben Noble was born on May 1, 1922, in a small fishing community in Fraserburgh, Aberdeenshire, Scotland. At the beginning of World War II he took the B.Sc. in radiophysics at the University of Aberdeen. He spent the rest of the war at the Admiralty Research Laboratory, doing experimental work in underwater acoustics. After World War II, he took the M.A. in mathematics at Cambridge. Instead of taking a Ph.D. he chose to join the Cambridge Mathematical Laboratory as an adviser on numerical methods. There he was influenced by the building of EDSAC, an early digital computer. This was followed by a period in industrial research with the Anglo-Iranian Oil Company (now BP) in England, from 1949 to 1952.

He then returned to the academy. From 1952 to 1955 he was first a research fellow and later lecturer at Keele University, Staffordshire, which had just started up as the first new university established in England since 1910. He then moved to the Royal College of Science and Technology (later, the University of Strathclyde) in Glasgow. There he took the D.Sc. (an English degree for independent research, without a supervisor) from the University of Aberdeen.

In 1962 he emigrated to the United States, where he held a visiting position at MRC from 1962 to 1964. From 1964 to 1985 he was professor of mathematics and computer sciences at the University of Wisconsin-Madison. This appointment included permanent membership in MRC from 1964, and he served as its director from 1975 to 1979. He became a citizen of the United States in 1975. He retired in 1985 with the rank of professor emeritus of mathematics and computer sciences, and now lives in England with his wife Denise.

During his career, Noble had broad research interests, including linear algebra, Fourier transforms, mixed boundary-value problems, integral equations, variational methods, computational methods, and (after he retired) multiple scaling and the numerical solution of antenna arrays. He has written four books and supervised fourteen students on various topics. He describes himself as "Jack of all trades, master of none."

Noble's visiting appointments during his professional career included posts at the Courant Institute, Oberlin College, Hertford College (Oxford), and the University of Lancaster, England. The distinction of which he is most proud is the U.S. Army's Decoration for Distinguished Civilian Service.

Externally directed program change

Around the end of 1976 the Army transferred Colonel Lothrop Mittenthal from the position of commander of ARO and chairman of the AMSC to an assignment in Europe. Dr. Jagdish Chandra assumed the position of chairman of the AMSC in addition to his duties as director of the Mathematics Division of ARO. Chandra was to play a central role in the Army's relationship to MRC from that point until the end of the center's existence. One of the significant points about his role was that, for administrative and organizational reasons, he was to exert significantly more direct influence over what happened at MRC than had any previous chair of AMSC or of its forerunner, the AMSG.

Noble's plan for areas of concentration

Noble had become director on August 1, 1975, at the conclusion of Creighton Buck's two-year tenure as acting director. In the Semiannual Report of March 26, 1976, he gave an overview of his thinking at that time on the ways in which the mathematics program at MRC should evolve. To begin with, he identified "areas of concentration" for the coming year in the following fields:[29]

- applied partial differential equations (which by the next report had become "evolution equations"),[30]
- mathematical programming and operations research,
- spline functions and approximation theory.

He also observed:

> It is tempting to think in terms of developing an area of concentration in the numerical solution of partial differential equations, partly to complement our efforts on the analytical side, partly because the subject is of great practical interest. . . . Experts in this area are in great demand, and in any case it is not clear that there is a U.W. tenure slot available even for the right person. The answer may be to encourage one of the MRC permanent members to specialize in this area.[31]

This observation seems prescient, given the later pressures on MRC to move more into computational mathematics and its ultimate failure to do so. It also points to a problem in the research relationship that may not be apparent at first glance, but that was very significant in the operation of MRC. That problem was the divergence in priorities between MRC on the one hand and the academic departments on which it depended for tenure appointments on the other. This divergence is so important that we deal with it later in a separate section.

The first peer review

One aspect of the increased supervision was the beginning of formal peer review of MRC's program. This was entirely new; there had never been any such review since the founding of MRC. Of course, the directors had reported semiannually to the AMSC, in considerable detail, about MRC's work, and they had made sure that the chair of AMSC was fully informed about their actions and plans.

The immediate occasion of the review was the submission of a new proposal for continued Army funding. The contract then in force ran until February 1980, and MRC intended to submit the final version of a proposal for continued funding by the autumn of 1978. This proposal would be subject to the normal external review by reviewers whom ARO would select. As a preliminary measure to develop information that MRC might use to prepare the strongest possible proposal, ARO commissioned an external peer review for the spring of 1977.

The following reviewers conducted the review on May 26–27, 1977:

- H. F. Weinberger (University of Minnesota, analysis),
- R. C. DiPrima (Rensselaer Polytechnic Institute, applied mathematics),
- H.-O. Kreiss (Courant Institute of Mathematical Sciences, New York University, and University of Uppsala, Sweden, numerical analysis),
- Herbert A. David (Iowa State University, statistics and probability),
- R. W. Cottle (Stanford University, operations research).

In addition, observers attended from the NSF, the ONR, and the Air Force Office of Scientific Research.[32]

Chandra summarized the results of the review in a letter to Noble dated August 30, 1977. Because of the importance to MRC of this review, and because the issues it raised were to plague the center from then until its demise, we extract parts of the letter here. Chandra had asked the reviewers to answer five questions, and we reproduce those in italics followed by the responses:

Is the present and projected emphasis in various subareas consistent with the main objective of the center?
"Through its present staff and visitors MRC has strength in the areas of nonlinear functional analysis, approximation theory and spline functions, mathematical programming and operations research, and statistics. All the panelists agreed that these areas fit within the objectives of the Center and should be supported. However, it is felt that there are more compelling areas such as physical and phenomenological mathematics and numerical methods for solutions of partial differential equations that would fit more closely with the objectives of MRC. Research in these areas is lacking; it should be developed as quickly as possible through the appointment of excellent senior faculty and should be the core interest of MRC. These research areas should be encouraged even if it requires curtailing other areas of research at MRC. These will be hard decisions that need to be made. Physical and numerical mathematics can have a synergistic interaction with the present strengths in nonlinear functional analysis and approximation theory and spline functions; and this interaction should tend to keep the latter more abstract areas potentially applicable."

Has the center been able to attract competent mathematicians? Is it able to attract competent mathematicians now, and what are the prospects for the future, in this regard?
"In the question of the Center being able to attract competent mathematicians, the consensus of opinions was that in the past MRC has been very successful. If the various departments at the University of Wisconsin (UW) continue to cooperate on joint appointments, the future looks bright for recruiting highly qualified mathematicians. At the present time there are several outstanding scientists with half-time appointments aboard. Some of these individuals could make valuable contributions to the work of MRC, and they would pose no commitments on the part of UW."

Are the resources of the center being used in the best possible way to attain the objectives of the center?
"The many ways in which the Center functions impressed the members of the reviewing committee. The interaction of the permanent members with the visitors and post-doctoral students is to be commended. The conferences and symposia along with the resulting published proceedings give evidence to the outside world of MRC's influence and effectiveness. A unique feature of the Center was signaled out for special praise. This was its ability to build up in [a] relatively short-time concentration of visiting experts to attack current research problems. An example being the year of concentration on evolution equations. Thus it was felt the resources of the Center have been used very effectively. It was also felt that the mathematical assistance and services that MRC makes available to the Army, has served to upgrade mathematical expertise within the Army laboratories. This assistance might well have been more effective had there been more concentration in physical mathematics and applicable numerical analysis. The recent addition of more statisticians to handle the many Army requests for assistance was praised."

The permanent members of the center form the core. How do you appraise their scientific stature, leadership, and productivity?
"Comments on this question, as was expected, depended heavily on the field of the reviewer. Professors Crandall, de Boor, Harris, Noble, and Robinson were particularly praised for being productive and in the forefront of their respective fields. Noble was singled out for his outstanding ability to apply mathematics to the real world problems. Obviously, more scientists of this caliber are needed at the Center."

Have you any suggestions to increase the effectiveness of MRC as a national center for applicable mathematics?

> "Several suggestions were made to increase the effectiveness of MRC as a national center for applicable mathematics. One that was emphasized by every panel member was the need for a change in the location of MRC. It would function better if it were placed near the libraries, and close to the mathematics and computer science facilities of UW. Obvious benefits would accrue to all concerned. The move should be made soon. I was pleased to learn that the University administration has the Center's relocation high on the list of its priorities. Certain branches of applicable mathematics should be added or emphasized to the list of the Center's main areas of concentration. Among these are numerical analysis, phenomenological mathematics, and stochastic modeling. The first two of these areas have already been mentioned earlier. It is quite possible that the funding might not be sufficient to excel in all areas stated in the objectives of the Center. It is desirable, however, to set priorities consistent with the overall objectives."

Chandra made a few more remarks praising the visitor program and Noble's effectiveness as director and noting the need to maintain a consulting capability in statistics. Near the end of the letter, he quoted one of the panelists on the role of MRC:

> It is a delicate give and take proposition for the Army to support a university academic unit that is dedicated to basic research and at the same time be responsive to immediate and short term Army needs. It seems to have worked reasonably well in the past and it may be possible to make it work better in the future, but possible conflicts of interest should be anticipated.

This letter is especially significant because it is the first open, written expression of dissatisfaction by the Army with aspects of the performance and direction of MRC. Behind the polite wording one can see three substantial issues:

- Physical and phenomenological mathematics and numerical methods for solution of partial differential equations are "more compelling" than MRC's current areas of strength, and work in these areas is "lacking." They should be built up quickly, even if other areas have to be cut back to do so; stochastic modeling also needs emphasis.
- Some of the permanent members are named favorably for their productivity; others are not.
- MRC, which since the bombing had been located on the western edge of campus in a modern building owned by the Wisconsin Alumni Research Foundation (WARF), should move back into the central campus area as quickly as possible.

In hindsight, one can better appreciate the quality of this review (and perhaps also its interpretation by ARO) by noting that each of these three problems was to recur from the time of the review until the end of MRC, and none was to be satisfactorily solved.

Even near the time of the review the difficulty of at least one of the problems was well understood at MRC. In the spring of 1978, not quite a year after the review had taken place, Noble reported to Chandra:[33]

> The location of MRC continues to cause concern. Although at present we have excellent quality accommodation, it is some distance from the departments and libraries. It seems impossible to move back to our previous ideal accommodation in Sterling Hall. . . . We have been reassured repeatedly since 1971 that our claims will receive full consideration as space becomes available in the central campus area. However we live in a democratic society where most decisions are taken by committee, and there is fierce competition for any space that becomes available. . . . I cannot at this point make any estimate of the probability that we will move back to a central campus location in the foreseeable future.

It was prudent of Noble to make no such estimate, because in fact MRC was never to leave the WARF building from then until the end of its life.

Organizational rigidities versus adaptation

Chandra's letter reporting the results of the first peer review had made it clear that the Army would require change in several aspects of MRC's operation. Some of the changes involved, such as a possible move back to the center of campus, did not really involve organizational questions, at least at the MRC level. The possibility of moving would depend on whether suitable space was available. The decision would also rest on whether the Madison campus administration considered a move by MRC to be desirable and of a sufficiently high priority to receive that available space in the face of competing claims by other campus activities. Desirability was not a trivial issue: because of the war-related political controversy and the ensuing bombing, some people on the campus vehemently opposed MRC's presence at all, and others were simply afraid to have it located near their workplaces.

On the other hand, some of the other issues, such as the buildup of physical and phenomenological mathematics, required changes in personnel. To accomplish such changes new people would have to be hired, and with static funding the resources allocated to some current staff would have to decrease. This posed little problem with respect to the visitor program, but with respect to the long-term staff ("permanent members") it raised very difficult issues involving both the nature of Madison campus faculty appointments and the relationships between MRC and various campus departments. This section explores some of those issues and shows how rigidities inherent in the faculty personnel system caused great difficulty for MRC in its effort to adapt quickly to environmental change.

The introduction of "regular members"

In the fall of 1978, Noble reported to Chandra on several organizational developments at MRC, two of which are of special significance for us. These were the designation of John A. Nohel of the Department of Mathematics at Madison as the next director of MRC, to take office on August 1, 1979, and a change in the system of permanent members.

Noble himself favored a rotation of incumbents in the director position, and the change of leadership was appropriately timed to fit with the expected commencement of the period of performance under a new contract, anticipated to run from December 1979 (just before the expiration of the existing contract) to 1984, with a possible extension to 1985. In fact, the new contract did not take effect until April 1, 1980, and it ran for a period of three years. The intent at the time was to have another peer review in 1981; on the basis of that review, ARO could decide to extend the period of the contract by a further three years.[34]

The change in the system of permanent members was more unusual. Ever since the inception of the MRC contract, the center had had permanent members. They helped to guide its research program and they generally were on the MRC payroll for significant portions of their time, although they could also teach courses for their campus departments. There had not been much change in the membership of this group, though H. F. Karreman had resigned his MRC appointment and moved full time to the School of Business at Madison in 1977. Now, for the first time, the system itself was to change. Noble explained:

> As you know, there have been significant changes in MRC during the last few years. Five years ago, MRC was run by a small number of "permanent members" who were supported essentially full-time on MRC funds. In order to broaden our program we have brought in members of other departments on a systematic and continuing basis. . . . The only way we have been able to afford this is that the permanent members have been teaching half-time, and the funds released in this way have been used to support the additional members of the MRC executive committee. . . . In the proposal we have gone one stage further by naming altogether 20 "regular members," a regular member being de-

> fined as "a mathematical scientist, normally having tenure in an academic department of the University of Wisconsin-Madison, who is expected to have a significant research (and/or administrative) role at MRC." . . . It is anticipated that, over the period of the contract, the regular members supported by MRC and the membership of the MRC Executive Committee will change in response to MRC's current programs and needs.[35]

This introduction of "regular members" was a significant organizational change. The idea had appeared as early as July 1975, when Creighton Buck had written to the then-chancellor of the Madison campus,

> I anticipate that it may be to the benefit of both the University and MRC if future "permanent" appointments to MRC are made with a five-year term, rather than indefinitely; this permits greater flexibility of program for MRC, and greater participation by members of the UW faculty, but also requires that the departments concerned modify their long range budgets accordingly. It is hoped that this will not hamper MRC in attracting new staff in those areas where it must now seek strength.[36]

Implementation of this idea represented a real effort by MRC to adapt to the new environment whose requirements were made plain by the spring 1977 peer review. Nevertheless, it was to prove insufficient.

One reason may have been the organizational structure on which it was predicated, namely, increased participation in the direction of MRC by tenured faculty of the Madison departments, who would be brought on board as "regular members" for varying periods of time depending upon MRC's programmatic needs. With the benefit of hindsight, one can see at least two difficulties with such an approach:

- The particular mix of skills and interests that MRC needed at any given time might or might not be represented among the tenured departmental faculty at that time. Even if they were, the faculty members concerned might or might not be available for release from their departments. If they were not so represented, the departments might not be willing to recruit in the areas of interest to MRC. We discuss this conflict of priorities in more detail in the following section.
- Faculty brought on board for temporary service as "regular members," and who had every reason to expect to be sent back to their departments the next time MRC's scientific priorities changed, would have very little reason to put MRC's welfare at the top of their own priorities.

The second point seems especially significant, and it points to an organizational dilemma that perhaps has no real solution, and that in any case MRC never succeeded in solving. First-class people will place an organization's welfare ahead of their immediate personal interests over a substantial time period only if they expect to have some long-term security in their relationship to the organization. But by offering such security, one locks the organization into supporting the specific areas and ways of thinking represented by those people, and this makes it very difficult to change direction quickly if the external environment demands such a change.

Other difficulties that MRC faced in reorienting its research program included the budgetary problems that the university itself faced in the early 1980s. Because appointees to MRC on other than a temporary basis had to have tenured appointments in Madison departments, problems in securing such appointments directly affected the center in its effort to adapt its research program. A good example was MRC's effort to hire a first-class senior person in continuum and solid mechanics. Nohel reported to Chandra in early 1981 that "the situation is further complicated by the fact that the Department of Mathematics has almost zero prospect of securing a senior-level position which could be joint with MRC in this vital area"[37] and went on to conclude that the best course of action for MRC would be to hire a junior person. That hire (M. Renardy) was in fact made, and proved to be very productive, but no junior faculty member could supply the mix of skills and visibility that MRC would have gained from a senior appointment.

Tenured appointments and departmental relationships

The conflict in staffing priorities between MRC and the academic departments stemmed from a very simple administrative problem. It would have been very difficult to get any first-rate person to come to MRC for an extended period (e.g., more than the usual year's leave) without a tenured appointment. This was true both because such people naturally worried about employment security and because without a regular faculty appointment the personnel system at Madison would have made such a person a second-class citizen. For example, as a member of the so-called "academic staff," such a person would have been able neither to direct graduate students nor to solicit extramural research funds without special permission. Further, the personnel system did not permit appointment at a regular senior faculty rank (professor or associate professor) without tenure. Therefore, in order for MRC to attract a senior person for an extended stay, it was necessary to arrange a tenured faculty appointment for that person.

However, tenured faculty appointments were not allowed in a soft-money operation like MRC; they were available only through the regular academic departments. Further, for such a department to appoint an individual to tenure meant that the department might have to live with that individual for the rest of the person's career, whether or not MRC support was available. Indeed, this was exactly what happened after MRC lost the Army contract. The tenured faculty with MRC appointments were picked up by their academic departments, as under the tenure agreement they had to be, whether or not those departments might have wanted to do so.

Under this system there was constant tension between MRC and the departments over long-term appointments. Visitor appointments, of course, involved no difficulty. They were made on contract money solely by MRC, except in the occasional case in which the visitor and an academic department came to a mutual agreement for the visitor to teach a course. Long-term appointments, however, required commitment of a tenured position, and the areas in which the departments wanted to grow over the long term were often quite different from those in which MRC wanted to appoint people. Therefore, the director of MRC was not entirely free to make appointments; he had to deal with the departments to get tenured positions. Moreover, commitments for such tenured positions could not be made by the department chair in quiet negotiations with the director but had to be approved by vote of the entire tenured faculty of the department, often by a supermajority. This put such questions squarely in the middle of departmental politics.

On the other hand, the director also had negotiating power because he could appoint regular faculty from the departments to the MRC staff for part-time research, thereby granting them much-desired time away from teaching. The departments themselves had no funds for such released time. The result was a constant negotiation process of academic horse-trading, in which both sides had to participate but neither side got all that it wanted. This difficulty had existed to some extent from the beginning of MRC, but it was considerably exacerbated in the period we will now discuss because MRC was subject to considerably more outside direction, and to the attendant changes in area emphasis, than it had been previously.

Funding shortages and shrinkage of the research program

By the spring of 1977, the areas of concentration had not changed, except that "mathematical programming and operations research" had become "nonlinear programming."[38] Within about a year, however, resource shortages forced MRC to reassess its plans for areas of concentration. In the spring of 1978, Noble wrote to Chandra:

> We have now had the experience of the evolution year, and of planning two areas of concentration for next year. There is no doubt in our minds that the evolution year has been a great success and that the idea of having different areas of concentration each year should be continued. However, the only way I have been able to afford substantial programs in

two areas, at the same time having a broad coverage of our other areas of strength, has been to use up reserves carried over from 1977–'78. I have come to the conclusion that unless additional funds are forthcoming, we should revert to *one* area of concentration in 1979–'80. There is general agreement that this should be physical mathematics.[39]

These funding difficulties reflected not only budget limitations but also inflation, which was becoming a major public policy problem in the Carter administration. This financial situation was not going to improve, as we will see in the final section of this chapter.

The last years of MRC

This section covers the end of the MRC operation, beginning with the second external peer review and ending with the Army site visit that effectively eliminated the Madison campus from further consideration in the competition for a new contract. It is relatively brief because this chapter has concentrated on organizational issues, rather than on research. Although the period we discuss here covers several years, there is little that is new in the organizational issues. The problems that we have seen before appear again, such as:

- *Organizational rigidity* versus the need to adapt the research program and the staff to changes in Army requirements and in the external scientific environment, in particular, the efforts to increase activity in physical mathematics and in what would now be called computational science.
- *Fiscal problems*, resulting from limited budgets and inflation, that squeezed the research program and eventually forced a shift in the mix of staff to more junior and fewer senior people.
- *The isolated location* on the west edge of the Madison campus and its negative consequences.

The second peer review

As noted earlier, a new contract for MRC had been negotiated with a start date of April 1, 1980, and a period of three years;[40] its extension for a second three-year period was to depend in part on the result of a second external peer review. This review took place slightly more than four years after the first, on September 14–15, 1981. The review committee consisted of:

- Julian D. Cole (UCLA),
- Herbert A. David (Iowa State University),
- James Glimm (Rockefeller University),
- George Herrmann (Stanford University),
- Peter D. Lax (New York University),
- Werner Rheinboldt (University of Pittsburgh),
- Hans F. Weinberger (University of Minnesota).

There were also twelve other observers and guests from various Army and other government agencies.

Chandra communicated a summary of the results of this review in a letter[41] to Nohel dated October 13, 1981. They included the following points.

- Physical mathematics, a new area at MRC, was "very important" and was "certainly growing in stature," although it had "not yet achieved the level of acclaim of the applied analysis group." He pointed out that the group in physical mathematics was not as cohesive as that in applied analysis and was "in urgent need of some additional member, perhaps a senior member in continuum mechanics."

- In numerical analysis, the committee was of the opinion that work in interval analysis should not get much further support.
- The effort in scientific computing should be increased.
- A predoctoral fellowship program should be considered.
- Work in operations research, except for some work in numerical optimization, should be phased out.

Receipt of this report led to additional efforts at MRC to reorient the program in order to demonstrate compliance. However, while these were in progress, some good news arrived in the form of a letter from Chandra dated November 13, 1981 announcing that the ARO Technical Review Panel had approved the Mathematics Division's recommendation to continue support of MRC for another three years (to spring 1986). ARO and the University of Wisconsin-Madison successfully completed the negotiations for that contract in December 1982, and the contract period was set at January 1, 1983 through March 31, 1986.[42]

Professor John A. Nohel

John A. Nohel was born in Prague, Czechoslovakia on October 24, 1924, the son of Arthur and Irene (Hönig) Nohel, and died in Zürich, Switzerland on November 1, 1999. He immigrated to the United States in 1939 and served as an enlisted man in the U.S. Navy during World War II. After the war he studied at The George Washington University in Washington, DC, where he took the B.E.E. degree in 1948. During this period he also met and married his first wife, Vera Weisskopf, with whom he lived until her death 40 years later in 1988. They had three children, Richard, Audrey, and Tom. In 1992 he married Liselotte Karrer, and they lived in Zürich until his death.

In 1953 Nohel completed the Ph.D. at MIT with a thesis directed by Norman Levinson. In the same year he joined the mathematics faculty at the Georgia Institute of Technology in Atlanta, where he stayed until 1961, rising to the rank of professor of mathematics.

In 1961 he accepted an offer to join the Department of Mathematics at the University of Wisconsin in Madison, where he remained until his retirement in 1991. At Madison he served as chair of the Department of Mathematics from 1968 to 1970, and as director of MRC from 1979 to 1987. After the termination of MRC in 1987 he accepted the directorship of the follow-on organization, the Center for the Mathematical Sciences, and served in that position until 1990.

Nohel's research was concentrated first in Volterra integrodifferential equations, and later in viscoelasticity and non-Newtonian fluid dynamics. He was coauthor or coeditor of twelve books and author or coauthor of more than eighty scientific papers. He was also a member-at-large of the Council of the Society for Industrial and Applied Mathematics from 1966 to 1969 and was a journal editor and reviewer. He held visiting appointments at the École Polytechnique Fédérale de Lausanne, Switzerland, the University of Paris, and the Helsinki University of Technology. His research and professional activity were recognized by his election to the grade of fellow of the American Association for the Advancement of Science in 1984.

Nohel was very active in organizations promoting human rights, including the Human Rights Committee for Mathematicians and Amnesty International. He had a special interest in helping young Czech mathematicians. He also had a very wide range of personal hobbies and interests, ranging from chamber music and opera to cross-country skiing and hiking.

Continuing fiscal problems

Although the new contract was in place, the fiscal problems affecting both universities and government research support in the early 1980s severely affected MRC's mode of operation. In early 1984, Nohel reported to Chandra that:

> Members of the AMSC are aware that the current ARO budget allows for no increase in funding during the calendar years 1983 through 1985. Because of this fact and in light of the long-term commitments, which MRC has made in order to implement the Peer Review recommendations, it is evident that MRC cannot at the same time maintain at previous levels the strong visitor program for which it is justly and internationally famous. From the appointments during the current year and from those projected for 1984–85, it is also evident that MRC is already appointing a much smaller number of very senior visitors for long periods of time (more than 2 months) unless they wish to visit at least partially on their own funds. Instead, we are successfully attracting outstanding postdoctoral researchers.[43]

Thus, the slow squeeze on resources, both at the University of Wisconsin-Madison (where in one year in the early 1980s faculty received no salary increase at all) and at ARO, was having its effect on MRC. However, the final act in the review and negotiation cycle was about to begin.

The end of the Wisconsin contract and the transition to Cornell

In September 1984, ARO issued a solicitation for a *Center of Excellence in the Mathematical Sciences* and invited proposals.[44] This was a fundamental change. For the first time since MRC had begun, a new contract was not being negotiated directly with the university, but rather the entire operation had been put out for bids. Moreover, as Nohel noted in a report to Chandra, MRC itself was to end. In Nohel's words, "the termination of MRC at the conclusion of the current ARO contract presents the University of Wisconsin-Madison with a clear challenge to select the best qualified interdisciplinary mathematical scientists for participation in the proposal in order to compete successfully for the new ARO-sponsored *Center of Excellence in the Mathematical Sciences*."[45]

Whatever was to come out of the new competition, it would not be the MRC operation that had existed for the past 28 years.

The Madison campus submitted a proposal for the new competition, with Nohel as principal investigator. Other institutions also submitted proposals, each of which received preliminary review by ARO. As the competition proceeded, finalists were scheduled for site visits by an ARO-organized team. The team, under the leadership of Jagdish Chandra, visited Madison on February 25–26, 1985.[46]

Some of the most important questions the team asked included aspects of cost sharing. In particular, as the Madison campus had had government investment for a long time in this area, they wanted to know what special consideration it might give to the new center, such as university-funded released time for faculty research. Other questions concerned plans that the university might have to develop collateral areas, such as increased computational power, that would support the contract activity if it were located at Madison. Such resource commitments were outside Nohel's scope of authority so they were handled by Irving Shain, then chancellor of the Madison campus, and E. David Cronon, then dean of the College of Letters and Science.

Shain recalls that he believed strongly that MRC belonged at Wisconsin, and he spoke to the representatives from that point of view. His general policy with regard to operations under the supervision of deans was that he would put in one-time money, but would not take on continuing commitments such as salaries, which would have to be paid for by the deans. However, he recalls that specific commitments of money on behalf of the center would have been difficult in any case at that time because the campus was in serious budgetary trouble due to difficulties with the state budget, as well as reallocation of funds from the campus to the University of Wisconsin system.[47]

Cronon, on the other hand, recalls that as dean he was not in a position to commit resources (such as released time for research) on behalf of the campus and that such decisions would have been made by Shain and by Robert M. Bock, then dean of the Graduate School. He also confirms Shain's recollection that the campus was in a period of particular financial stress because of shortfalls in anticipated state funds as well as actions by the University of Wisconsin system, and notes that resource commitments such as released time would have caused significant problems to the college. In such an environment he was very sensitive to any demands on the instructional budget, and there simply was no money in that budget to use for extra commitments such as released time.[48]

In fact, Cronon's impression was that the Army representatives had already made up their minds to move the contract away from Madison, and that the visit was more a *pro forma* exercise than a real review. He did try to assure the Army representatives that MRC was very valuable to the Madison campus and to point out to them how much Madison had contributed to the Department of Defense through MRC's work over a period of many years.[49]

Communication between Cronon, who was a historian by training, and the site visitors may have been hindered by their very different professional backgrounds. One such case occurred when one of the representatives asked what the College of Letters and Science proposed to do to increase computing power on the Madison campus. Cronon replied that a high priority of the college was to ensure that every faculty member had a PC on his or her desk. Although this answer was absolutely accurate—it was a high priority at the time, and faculty in many Letters and Science departments then lacked access to PCs—it was not the sort of answer that the team was expecting.

The net result of the team's discussions with Shain and Cronon was that the Madison campus would not offer any incentives, such as released time, of the kind that the site visit team was looking for. When this became clear, Chandra thought that perhaps the campus administrators had been insufficiently briefed on what topics the site visitors would raise. He therefore suggested recessing the meeting for an early dinner, after which they could continue the discussion, having in mind that a recess might give Nohel a chance to strategize with the administrators. When the meeting recommenced, it became clear that Madison's response was still that it wanted the contract but would do nothing out of the ordinary to get it. There would be no released time, nor any other special incentives.[50] At that point the site visit terminated, and shortly thereafter, ARO notified Madison that it had been eliminated from further consideration in the competition. The award eventually went to Cornell University.

The contract then in force for MRC was scheduled to terminate on March 31, 1986. Subsequently the university negotiated a no-cost extension of that contract, with some reprogramming of funds, through January 31, 1987.[51] As the original contract had been signed on April 25, 1956,[52] MRC had lasted for nearly 31 years. During that time it had produced nearly 3,000 Technical Summary Reports, many of which later evolved into journal publications, as well as the proceedings of its regular semiannual conferences. For a substantial part of its existence, it was a world leader in important areas of applied mathematics. By academic standards, this is not a very long run, but in the realm of contract research it was a very creditable performance.

Conclusion

The 1970s witnessed stresses on various scientific organizations. In particular, conditions for the support of research in mathematics—nationally, within the Army, and at the Madison campus—had substantially changed. These changes presented MRC with unusual challenges. The most significant of these was the introduction of considerable flexibility into its program and the adaptation of that program to changes in the external research environment. The new external pressures forced the center to reexamine some of the structural rigidities of its organization, such as the institution of permanent members, and some of its accustomed ways of organizing and conducting research

A major source of externally driven program change was the start of a peer review process instituted by ARO. It brought to the fore many of the inherent difficulties in long-term support of centers like MRC. To quote an early peer panelist, "It is a delicate give and take for the Army to support a university academic unit that is dedicated to basic research and at the same time be responsive to immediate and short term Army needs. It seems to have worked reasonably well in the past and it may be possible to make it work better in the future, but possible conflicts of interest should be anticipated." For instance, the particular mix of skills and interests that MRC needed at any given time to respond to the anticipated program changes or evolution might or might not be available through its pool of permanent members or the extended family of tenured (related) departmental faculty. A solution to this problem was sought by bringing on board for fixed periods of time faculty members or visitors, who would have very little reason to put MRC's interest at the top of their own priorities. MRC was never able to resolve satisfactorily this conflict between its staffing priorities and those of the academic departments on the Madison campus.

While such issues preoccupied the leadership of the center, MRC nevertheless achieved many outstanding technical successes during this period. The years of concentration, and their logical evolution and transition, resulted in significant advances in the areas of nonlinear analysis, continuum mechanics, and reactive flows. In particular, under the leadership of Crandall and Nohel MRC was able to assemble a very strong group of researchers in time-dependent partial differential equations, perhaps the strongest group anywhere in the world at that time.

Chapter 4: An Intellectual Crossroads

Europe has nothing like the MRC, where scientists are given complete freedom to think, with no lectures or administrative responsibilities—no interruptions.
—Alexander Ostrowski, Basel, Switzerland, 1961

Introduction

At a time when the ascendant view of mathematics was that of pure mathematics (much of it in the Bourbaki tradition), the Mathematics Research Center presented a more complete view of the mathematical world. MRC provided a rich intellectual environment where resident mathematicians continuously interacted with a stream of visitors both domestic and foreign. In a very real sense it was an intellectual crossroads in the world of applied mathematics, and one could argue that for a considerable time it was preeminent in the world in that respect.

Many senior members of the center had a rich history of working on applied problems, generated either by other disciplines or by real-life applications. Several bright young scholars who visited the center or were postdoctoral fellows there benefited immensely in the formulation of their perspective on the world of mathematics and future directions and opportunities for creative mathematical research.

A mathematician may obtain great benefit by solving a practical problem, and the center provided ample opportunities for such solution through its interactions with the Army and other government agencies. However, MRC provided more than that: namely, an opportunity to pursue related questions that were, at the core, of purely mathematical interest. This allowance to deviate on mathematical tangents was at the crux of the center's intellectual achievements. In this chapter, we describe a few of these outstanding scientific accomplishments, which have had a profound impact on application domains as well as on the evolution of applied mathematics during these decades. We also mention other aspects of the MRC environment that were important in promoting and sustaining this scientific productivity.

Spline functions

Background

Spline functions are a class of piecewise polynomial functions satisfying continuity properties only slightly less stringent than those of polynomials. They are a natural generalization of polynomials, and they have highly desirable characteristics for use as approximating, interpolating, and curve-fitting functions.

It was Newton who pioneered the use of polynomials for obtaining numerical approximations for interpolation, integration, differentiation, and other such operations on functions of one variable. In these methods, interpolation plays a preferred and fundamental role. In 1949, Arthur Sard introduced a new point of view leading to his best approximation formulae. In this new approach to Newton's old problems, it seemed as if interpolation was losing the fundamental role that it played in Newton's work. It turns out that interpolation can be restored to its central role, provided that we replace the Newtonian polynomial interpolation of the data by interpolation using the so-called natural spline functions.

Early history

Although they were not referred to by this term, spline functions had been used before mid-century in a few isolated instances. In 1904, Carl Runge used periodic quadratic splines. In 1938, Collatz and Quade used periodic splines of degree *n*, for the purpose of approximating the Fourier coefficients of empirically defined periodic functions. Actuarial mathematicians used osculatory interpolation functions for data fitting.

I. J. Schoenberg was the first to name and single out these functions for special study in the mid-1940s. Schoenberg was a consultant at the U.S. Army Ballistics Research Laboratory (BRL) during this period. He was confronted with a practical problem. From actual firings of projectiles of different sizes and shapes, about two dozen tables of drag coefficients were to be constructed. These were then used to produce the trajectories of the projectiles, by integrating the differential equations that describe the motion. The calculations were done by hand on a desk calculator.

At that time, the first electronic digital computer, ENIAC, was developed at the Moore School at the University of Pennsylvania. One of its intended uses was the computation of trajectories. For the integration method used on ENIAC, the available tables of the drag coefficients were too rough. These tables required smoothing or graduation. More precisely, the empirical functions described by these tables had to be approximated by very smooth functions having very smooth first and second derivatives. These approximations were to be computed to eight significant figures, with the first and second derivatives also computed to eight significant places. Due to the complicated trend of the curves representing drag coefficients, traditional methods of approximation proved to be of no avail. Schoenberg had a perfect motivation for his work on spline functions.

What are splines?

The simplest functions in pure and applied mathematics are polynomials. For example, if we let π_k denote the class of polynomials of degree not exceeding *k*, then there are

- π_0, the constants,
- π_1, the affine (linear) functions,
- π_2, the quadratics.

High school students learn about these in basic mathematics courses, and sometimes about the more intriguing family of cubics, π_3, and so on. A fundamental theorem due to Weierstrass asserts that polynomials will approximately imitate within a prescribed tolerance any kind of continuous curve in a compact range. This is a theoretical foundation for the use of polynomials.

Beyond polynomials, of course, we have step functions, which are clearly related to π_0. If we take the integral of a step function, we obtain a piecewise linear (broken-line) function. If we integrate again, we get a succession of parabolic arcs. Notice that the order of continuity increases with each integration. We already have the beginning of spline functions. Thus, a spline function of degree *m* (S_m) can be defined as a succession of polynomial arcs of degree not exceeding *m*, which are joined at certain points called *knots*, at which the order of continuity is as high as it can be without all of the arcs' being part of the same polynomial. A spline function (S_m) differs from a polynomial (P_m) of degree *m* in that both have *m* derivatives, but while the *m*th derivative of a polynomial is a constant, the *m*th derivative of a spline function is a step function.

Professor Isaac J. Schoenberg

Isaac Jacob Schoenberg was born in Galatz (Galați), Romania in 1903. He received a M.A. from the University of Jassy (Iași), Romania, in 1922. Schoenberg continued his studies at Göttingen and Berlin from 1922 to 1925, and he received his Ph.D. in mathematics from Jassy in 1926. He came to the United States in 1930 and served in various postdoctoral capacities at the University of Chicago and Harvard. Schoenberg was a member of the Institute for Advanced Studies at Princeton from 1933 to 1935. He went on to Swarthmore College and Colby College and later served on the faculty of the University of Pennsylvania from 1941 to 1965. It was during this period that he served as a research mathematician at the U.S. Army Ballistics Research Laboratory

(BRL). His pioneering contributions to the theory of spline functions can be traced to his time at BRL. In 1965, Schoenberg joined the University of Wisconsin as a professor of mathematics and as a permanent member of the Mathematics Research Center (MRC). In 1977, he became professor emeritus and embarked upon a series of visiting appointments at other institutions, including the Universities of California (San Diego), Florida, and Pittsburgh, the Weizmann Institute of Science in Rehovot, Israel, and the United States Military Academy at West Point, NY. He died in 1990.

Schoenberg made outstanding contributions in many areas of mathematics. His early work pioneered and elaborated the basic concepts of total positivity and variation-diminishing transformations. Total positivity is a concept of considerable power that plays an important role in various domains of mathematics, statistics, and mechanics. Another of his important mathematical contributions was to the area of finite distance geometry, where Schoenberg extended and refined earlier fundamental work of Menger and Blumenthal. His joint work with von Neumann made significant progress in the field of metric geometry of infinite-dimensional Euclidean spaces. Schoenberg was a leader in offering fundamental and far-reaching extensions of the concepts of positive definite and completely monotone functions defined on general metric spaces.

However, he is best known by far for his work on, and his realization of the importance of, spline functions for general mathematical analysis and approximation theory. The fundamental papers by Schoenberg on this subject gained him international renown, and they stand as landmarks in the history of that subject as well as in its conception and development.

Schoenberg was also a man of broad culture. He was fluent in several languages and a great lover of art, music, and world literature. Affectionately known to his friends as Iso, he was a very gracious and giving individual who made deep and enriching contributions to the domain of the mathematical sciences.

Splines and MRC

Spline functions made their first appearance at MRC quite early, namely, in the first MRC Symposium on Numerical Approximation that was organized in 1958 by Langer. Although the symposium volume contains a paper by Schoenberg, the major spline paper presented was by Michael Golomb and Hans Weinberger on "Optimal Approximation." The paper proves results about the variational approach to splines that were eventually redone by others over the next twenty years. The reason for this was that even the printed version of the Golomb–Weinberger contribution fails to use the term "spline." This is all the more surprising, given Schoenberg's presence at the meeting. This may be explained by character traits of Schoenberg, who was completely absorbed in mathematics and did not want to be disturbed by proprietary claims. What is important for our purposes is that the fundamental paper by Golomb and Weinberger was the result of their conversations as visitors at MRC. This is a typical example of the beneficial effects that MRC had in bringing professionals together and sparking innovative discussions.

Schoenberg's paper at this symposium dealt with variation-diminishing methods. It did not explicitly mention splines and B-splines, even though the best-known current variation-diminishing approximation methods are all spline-based, including Schoenberg's variation-diminishing method developed at MRC in the late 1960s. This also gave a student, Martin Marsden, who attended a class on approximation theory taught by Schoenberg, the opportunity to prove what is now called the Marsden identity, a basic result about the representation of polynomials as weighted sums of B-splines.

Schoenberg's first paper on splines written as an MRC member was joint with Curry under the title "Polya Frequency Functions IV, the Fundamental Spline Functions and Their Limits." It has become the basic historical reference concerning B-splines. The major spline-related topics that Schoenberg, Carl de Boor, and others initiated at MRC are the following:

- variation-diminishing splines, including what is now known as Schoenberg's variation-diminishing spline approximation;

- cardinal splines, as eventually summarized in Schoenberg's SIAM lectures on the topic. This theory also provided much impetus to multivariate approximation, as it is a very good and thoroughly worked out example of current approximation from shift-invariant spaces;
- the Kolmogorov–Landau inequality.

While some of de Boor's work on splines was done before he came to MRC (such as recurrence relations for B-splines, including the FORTRAN package for calculating splines via B-splines), these important results were recorded in early MRC technical reports. Areas of spline function research that had their origin at MRC or were strongly developed there included cardinal splines, discrete splines, variation-diminishing spline approximation (smoothing splines), and multivariate B-splines of various sorts. Much current work on multivariate approximation, including some work on wavelets, had its start at MRC.

Some applications

The use of B-splines and total positivity in CAD/CAM (computer-aided design/manufacturing) is a nice example of seemingly theoretical mathematical work giving a very effective answer to an eminently practical problem. Total positivity is a matrix property. A matrix is totally positive if all of its minors are nonnegative. Schoenberg introduced this property in the 1930s, in a study of refinements of Descartes's rule of signs. It was picked up by the great Soviet mathematician M. G. Krein as a tool for the study of small vibrations of oscillating mechanical systems. In the United States, Samuel Karlin eventually became interested through Schoenberg's efforts, and he has since demonstrated the wide applicability of this idea in statistics and biology.

B-splines are particularly convenient building blocks for making up a spline. While this was recognized in the 1940s in the work by Schoenberg and Curry, it became a practical reality with the stable recurrence relations and the evaluation algorithms developed in the 1970s by de Boor.

A basic problem of CAD/CAM is the construction of a smooth curve (or surface) to a designer's specification. "Construction" is defined as the development of a mathematical formula, which can be evaluated to give any particular point on the curve or surface for purposes of plotting, of composing complex objects from simple ones, or of machining the surface with the aid of automatic milling machines, to cite some examples. The design takes place interactively. A rough outline of the curve is made, typically just a sequence of points in the plane, and it is left to a program to come up with a smooth curve that looks like the sketch. This curve is then modified, locally or globally, until it fits the designer's original concept. Local flexibility and faithful reproduction of the overall proposed shape are the main requirements of the mathematical curve description used in such programs. Splines, particularly when built up from B-splines, provide both. Local flexibility (and computational efficiency) derives from the fact that each individual B-spline matters only for a small part of the interval of construction. Faithful reproduction of the overall shape comes from the mathematical fact that the matrix that contains the values of all of the B-splines in question at an increasing sequence of points is totally positive. This total positivity implies that the spline curve's shape (its areas of concavity, of convexity, monotonicity, and the like) mimics that of a polygon whose vertices are, in effect, given by the B-spline coefficients. This means that the spline curve's shape can be prescribed by controlling the shape of the polygon (e.g., the B-spline coefficient sequence). Thus, one works with a very simple object (a polygon) yet obtains a smooth curve (the spline). More than that, total positivity makes it easy to derive a second polygon so that the actual spline curve lies between these two.

Schoenberg's variation-diminishing spline approximation scheme mentioned earlier exploits this relationship between the polygon derived from the B-spline coefficient sequence and the spline itself. Schoenberg's scheme was introduced into CAD by William Gordon around 1970 and immediately caught on, since it provided much-needed local flexibility, which earlier polynomial-based CAD schemes did not have. Gordon's student, Richard Riesenfeld, contributed much to spreading the idea through the CAD community. Parallel efforts at BRL have also led to some interesting solid modeling and computational geometry tools with a wide range of applications.

In another direction, Jeffrey Lane, a student of Riesenfeld, looked for simple proof of the total positivity property of the relevant matrices. He found a nice geometric argument, in which he described a fixed spline in terms of more B-splines. This means that the corresponding polygons become increasingly fine, until eventually they become indistinguishable from the spline curve. This leads to a useful scheme for evaluating B-spline curves. One obtains a polygon, which is the only kind of curve standard drafting equipment can draw, but it is an efficient polygon (e.g., a polygon which approximates the spline curve to within graphing accuracy but no more). These polygons had been identified and studied earlier by de Boor as a kind of "discrete spline." Lane, Lyche, and Riesenfeld made use of this identification to develop the *Oslo algorithm*, in analogy with the stable evaluation algorithms for continuous splines.

This ability to move quickly from the smooth spline to an approximating polygon (or an enclosing pair of polygons) becomes particularly important in surface work. When constructing complex objects from simple ones, it is important to worry about surface intersections. It is much easier to determine the intersections of polyhedral surfaces than those of the spline surfaces themselves. Thus, one obtains good first approximations using polyhedral surfaces (known to envelop the actual surfaces) and refines the approximations to the intersections as the polyhedral surfaces are developed.

Impact

Work at MRC, especially that of Schoenberg, de Boor, and many of their students and associates, has been recognized worldwide for its contributions and realization of the importance of spline functions for general mathematical analysis and approximation theory, their key relevance in numerical procedures for solving differential equations with initial and/or boundary conditions, their use in the construction and analysis of multiscale and multiresolution methods (e.g., wavelets), and their role in the solution of a whole host of variational problems. The fundamental papers by Schoenberg and de Boor[53] not only are critical to the history and evolution of the subject but also reflect its germination. Specifically, Schoenberg established[54] the intrinsic connections between monosplines and best quadrature formulas and determined their complete characterization. By introducing a new class of splines (namely, the one-sided Euler splines), he obtained sharper bounds on the higher derivatives of functions, a problem that has engaged mathematicians of the caliber of Landau, Kolmogorov, and others. Along with de Boor, Schoenberg developed a comprehensive theory of cardinal splines and their wide ramification in approximation theory.

The subject of spline theory and its applications has mushroomed, yielding results in many domains, including generalized interpolation theory, Sobolev spaces for qualitative and quantitative theory of differential equations, numerical methods in general, statistical estimation and regression analysis procedures, CAD/CAM, control theory, and multiresolution analysis of images and patterns. The current work on multivariate splines, which has tremendous potential for multidimensional signal processing and image analysis, builds on the foundation and successes achieved in the pioneering work at MRC.

Viscosity solutions

Background

Equations of Hamilton–Jacobi (HJ) type arise in many areas of applications such as optimal control theory, differential games, variational problems in physics, and mathematical finance. Specifically, consider problems involving HJ equations, which we take to be either the stationary form $H(x, u, Du) = 0$ or of the evolution form $u_t + H(x, u, Du) = 0$, where Du is the spatial gradient of u. Classical analysis of associated problems under boundary (Dirichlet) and/or initial (Cauchy) conditions by the method of characteristics is limited to local application owing to the crossing of characteristics. More precisely, if H and the initial conditions are smooth, the associated initial value problem will typically have a unique smooth (classical) solution over some maximal time interval, at the end of which Du becomes discontinuous (or develops shocks). Nonetheless, in fields such as control theory there is a globally well-defined function, the value function, associated with the problem. In the 1960s and 1970s researchers such as A. Douglis, W. H. Fleming, and S. N. Kruzkov in Russia had proposed generalized notions of solutions, which partly ameliorated this situation. Their results concerned equations that were convex in the gradient, a condition satisfied by HJ equations arising from control theory. Another approach, known as "nonsmooth analysis," which developed around the work of R. T. Rockafellar and his former student F. H. Clarke, had other ways to approach nonsmooth solutions in this control theory setting.

None of these approaches addressed equations that are not convex in the gradient. For instance, equations arising from differential games do not enjoy convexity. Moreover, these approaches do not readily adapt to second-order equations, which arise from stochastic control theory and stochastic differential game theory. In addition, fully nonlinear partial differential equations (the terminology is explained in the next paragraph) arise also in differential geometry, image processing, and other fields. Finally, the control theory of partial differential equations leads to HJ equations set in infinite dimensional spaces.

What are viscosity solutions?

It is easiest to explain the notion of a viscosity solution, as it is most often used today, in terms of second-order equations. A general second-order scalar partial differential equation can be written in the form $H(x, u, Du, D^2u) = 0$. Here u is the unknown real-valued function, Du is its gradient, and D^2u is the Hessian matrix of all second-order partial derivatives of u. The defining function $H(x, r, p, X)$ depends on the n-vector x of independent variables (one of which can be time), a single variable r for the "u slot," another n-vector p for the gradient slot, and an $n \times n$ matrix X for the D^2u slot. This equation admits a formal maximum principle provided it is monotone in X (in the usual ordering on symmetric matrices). We take it to be monotone nonincreasing. Likewise, monotonicity in r is natural here, so the net structural assumption we make is $H(x, r, p, X) \leq H(x, s, p, Y)$ provided that $r \leq s$ and $X \geq Y$. In this event, if u is a smooth solution of $H(x, u, Du, D^2u) \leq 0$ (i.e., u is a *subsolution* of the equation and ϕ is smooth and z is a local maximum of the difference $u - \phi$, then by calculus one has $Du(z) = D\phi(z)$ and $D^2u(z) \leq D^2\phi(z)$, so that the inequality $H(z, u(z), D\phi(z), D^2\phi(z)) \leq 0$ follows from the monotonicity of H in its matrix argument.

The inequality $H(z, u(z), D\phi(z), D^2\phi(z)) \leq 0$ does not involve derivatives of u, and this is the key. Requiring this inequality at local maximum points of the difference $u - \phi$ for every "test function" ϕ defines the notion of *viscosity subsolution*. Similarly, one defines a viscosity supersolution, and then takes a *viscosity solution* to be a function that is both a viscosity subsolution and a viscosity supersolution.

Note that the equations $H(x, u, Du, D^2u) = 0$ discussed earlier do not require H to be linear in any of its variables; the equation is *fully nonlinear*. Moreover, the viscosity solution notions do not require H to depend on D^2u; it may be a first-order differential operator. In fact, in this lies the origin of the terminology *viscosity solution*. A first-order equation $H(x, u, Du) = 0$ can be regarded as the limit of second-order equations $H(x, u, Du) - \varepsilon \Delta u = 0$ as $\varepsilon \downarrow 0$; this is the method of "vanishing viscosity."

Viscosity solutions at MRC

In a seminal paper,[55] Michael G. Crandall (a permanent MRC member) and Pierre-Louis Lions (a junior visitor at the center), using a notion they called viscosity solutions that is equivalent to but slightly more complex than that explained above, proved the first general global existence and uniqueness theorems for nonconvex first-order equations $H(x, u, Du)=0$. The value of the concept is established by the fact that very general existence, uniqueness, and continuous dependence results hold for viscosity solutions of many problems encountered in several fields of applications. For instance, for a large class of optimal control problems, the value function is the unique viscosity solution of the related dynamic programming equation. In the case of a deterministic problem, this equation is a first-order partial differential equation, usually referred to as the Hamilton–Jacobi–Bellman (HJB) equation, and for a controlled diffusion process,[56] the dynamic programming equation is a second-order parabolic differential equation. As viscosity solutions need not be differentiable anywhere, they are not sensitive to the classical problem of the crossing of characteristics.

The notion of a viscosity solution admits several equivalent formulations.[57] In fact, L. C. Evans, a former student of Crandall partly trained at MRC, had used the technique of putting derivatives on test functions to study some problems for fully nonlinear equations. In TSR 2390 Crandall, Evans, and Lions showed how to recast the original proofs with the definitions given above. Crandall and Lions also initiated approximate procedures for the construction of viscosity solutions. Specifically, they established the convergence of a class of difference approximations to these solutions by obtaining explicit error estimates.[58] Another student, P. Souganidis, did further work on approximation in TSR 2511. Also, at MRC Crandall and Lions initiated the theory of viscosity solutions in infinite dimensions, corresponding to control and differential game problems involving partial differential equations. They took up other topics as well, some with other visitors to MRC.

The extensions to second-order equations took some time. Lions provided some first results for equations convex in the Hessian. Later, R. Jensen made a breakthrough in treating nonconvex equations. Jensen, Lions, Ishii, Souganidis, and Crandall brought the field to a relatively high state in various contributions. The results of these activities and contributions by others from abroad are summarized in a very influential paper entitled "A User's Guide to Viscosity Solutions."[59] This work was done after the closure of MRC, but with support from ARO.

Applications

The theory and applications of viscosity solutions have been an active area of research. For instance, an abstract (operator theoretic) formulation captures situations where dynamic programming equations are not necessarily differential equations. For instance, the optimal control of a Markov chain yields a difference equation, and a piecewise deterministic process gives rise to a system of first-order partial differential equations. The theory of viscosity solutions is not limited to dynamic programming equations. Indeed, as noted, this theory applies to any equation with a certain maximum principle.[60] There is an obvious, intimate relationship to the theory of differential inequalities and to the construction of sub- and supersolutions.

The scope of viscosity solutions has subsequently been extended to a broad class of mathematical problems ranging from modeling and analysis of optimal and stochastic control to phase transitions, image processing, variational problems in physics, derivation of transport equations, and turbulent combustion. For example, another important application of viscosity solutions is to the theory of differential games. Specifically, the theory of two-person, zero-sum differential games started at the beginning of the 1960s with the pioneering works of Isaacs in the United States and of Pontryagin and his school in the Soviet Union. The principal motivation of these efforts was applications to military problems such as pursuit-evasion games. Isaacs based his work on the dynamic programming method. He analyzed many special cases of HJ equations, trying to obtain explicit solutions. This enabled him to synthesize optimal feedback from these explicit solutions. However, in general, one encounters singular surfaces, which are essentially the sets where the value function is discontinuous or not differentiable. The notion of viscosity solution provides an excellent way to avoid this problem in a large class of such problems. Evans and Souganidis, in TSR 2492, first verified that value functions arising from differential games are viscosity solutions.

The theory and applications of viscosity solutions is a most exciting success story in the field of mathematics. A clear indication of this is that since its introduction in 1981 by Crandall and his then young associate Lions, more than a thousand papers relating to the subject have been published in journals devoted to both mathematics and various application domains. For instance, these generalized notions of solution have solved many difficult problems of optimal control of diffusion processes with jumps and related HJB equations. Another, rather surprising area of application is finance. Mathematical finance is a relatively new area where viscosity solutions are likely to have a significant impact. Topics such as asset pricing models with stochastic volatility and transaction costs and investment problems involving jump processes directly benefit from the advances in this area. Some of the current research topics that make use of these solution notions include investment-consumption models, incomplete markets, and economic equilibrium theories.

The wide range of applications that can be studied using viscosity solutions has generated an enormous interest in the development of numerical methods and computational algorithms for the solution of such problems. For example, there is considerable research activity dedicated to the development of numerical schemes for the solution of HJB equations and their implementations on high performance computing platforms. Another active area is numerical methods for computing the prices of various exotic options in financial markets and the development of optimal strategies in portfolio management.

A proof of the importance and relevance of viscosity solutions in applied mathematics is the fact that they have already been used successfully to analyze a large number of questions related to the applications mentioned above. Lions received the Fields Medal in 1994 partly because of his contributions to this field. Similarly, Crandall was awarded the Steele Prize for Seminal Research by the American Mathematical Society partly for his work on viscosity solutions. Interest in this area remains robust and the method is reaching further into varied fields. Many of the people carrying out this work were former students at, or visitors to, MRC.

Generalized equations in nonlinear programming

In the late 1960s the field of nonlinear programming was still in a state of intensive development. One of the areas of investigation was the proper way in which to include the effects of feasibility constraints in the mathematical analysis of these problems. These constraints were typically of one or both of two forms: equations of the form $f(x) = 0$, or inequalities of the form $f(x) \leq 0$, where in either case the function f is a linear or nonlinear function from $\mathbf{R}^n$ to $\mathbf{R}^m$, and the inequalities are to be interpreted componentwise.

A great many nonlinear programming problems can be written with feasible sets of the form $\{x \mid g(x) \leq 0, h(x) = 0\}$, where the functions g and h, from $\mathbf{R}^n$ to $\mathbf{R}^p$ and $\mathbf{R}^q$ respectively, are actually fairly smooth (typically, at least C^2). The presence of continuous derivatives made it seem plausible that one should be able to apply calculus methods to questions involving these problems, and of course this had been done for many years with equality-constrained problems. However, it was not clear whether the same methods were appropriate for use in problems containing inequality constraints.

Workers in the field recognized that the presence of inequality constraints introduced a fundamentally new dimension into the problem of analysis, although the nature of this change was not well understood at the time. Some differences, however, were clear. For example, the well-known separation theorems that provided the main tool for deriving optimality conditions associated inequalities with sign-constrained dual variables, but equations with unconstrained variables. The same distinction arose in a superficially different, but actually identical, way in the duality theory of linear programming. However, most of the analysis done to investigate questions such as existence and stability of solutions relied on the traditional methods of calculus, which did not make specific allowance for inequalities. A notable, but somewhat isolated, exception was the constraint qualification introduced in 1967 by Mangasarian and Fromovitz,[61] which treated the inequality and equality constraints differently, and moreover in which the differences appeared directly analogous to those found in the known methods of separation and linear-programming duality. An open question was therefore whether one could develop general analytical methods for nonlinear programming problems that exploited the differences between equations and inequalities in a systematic way, so that it would be clear how to apply the resulting methods to any combination of constraints found in a particular problem.

Robinson began to work on this problem in 1970, while writing his Ph.D. dissertation under the direction of O. L. Mangasarian. He hypothesized that there should be a natural way in which to treat the entire expression defining the feasible set as a kind of operator, and that if the right kind of calculus to handle such operators could be found, then the problem of analytical methods for dealing with feasible sets would be solved. In hindsight, one can see clearly in this basic approach the influence of the operator-theoretic approach to nonlinear functional analysis then being stressed by many investigators at MRC. Indeed, one influence that contributed strongly to Robinson's choice of this approach was a course entitled Numerical Functional Analysis that he took in 1969–1970 from J. W. Daniel, who was then on the faculty of MRC and the Computer Sciences Department and who had originally come to Wisconsin as a postdoctoral research appointee at MRC. Other influences also contributed, including the weekly MRC colloquium lectures, many of which were on topics in nonlinear analysis (see the next section).

By 1971 Robinson had developed an initial approach to the inequality problem, based on writing the constraints in the form $0 \in f(x) + K$, where K was a closed convex cone. This method had the advantage of making it possible to write many different kinds of feasible sets in a uniform way: for instance, in the example given above one takes K to be the product of the origin in $\mathbf{R}^p$ and the nonnegative orthant in $\mathbf{R}^q$. This approach produced some initial publications.[62,63] However, it had an unsatisfactory aspect, which can be briefly described by saying that it was suitable when the value of $-f(x)$ happened to be at the vertex of the cone K but was too strong otherwise. He was able to remove this difficulty over the next two to three years, and the papers[64,65] describing the resulting analytical approach for linear and nonlinear inequalities have since been widely cited.

Although the inequality problem was important and the results useful, its solution represented a prelude to the establishment of similar results for the solutions of optimization and equilibrium problems themselves (as opposed to the sets of feasible points, to which the inequality work applied). Robinson believed that similar operator-theoretic methods could be found for that application too, but he deferred that work while completing the machinery for inequalities. In the middle 1970s he discussed his plans for working on this problem in a conversation with Jerzy Łoś, an eminent Polish mathematician then visiting MRC as part of the mathematical economics program. Łoś cautioned him, "Don't wait too long to do it, or someone else will get there first."

Perhaps motivated by this remark, Robinson had developed within two years an approach to analysis of the entire problem in which the closed convex cone found in the inequality analysis was replaced by a normal-cone operator (an instance of a *maximal monotone operator,* one of the roots of whose development incidentally had been an MRC TSR written much earlier by E. Zarantonello[66]). The paper[67] describing this "generalized equation" and the methods for its analysis has since been extensively cited, and the regularity condition it introduced has become the standard condition used for analysis of such problems.

The introduction of generalized equations and the associated methods of analysis was one of the principal accomplishments cited in the award to Robinson, nearly twenty years later, of the George B. Dantzig Prize of the Mathematical Programming Society and the Society for Industrial and Applied Mathematics. The prize citation did not mention the stimulating environment of MRC, whose influence can be clearly seen in the methods and approach that led to those results. Indeed, by that time MRC no longer existed.

Applied functional analysis

The influence of MRC in spreading the use of functional analytical methods in the development of mathematical programming, optimization, variational methods, approximation theory (including spline functions), and in general numerical methods was a hallmark of this period. Functional analysis was one of the most significant mathematical developments of the early twentieth century. This striking change of the framework in which mathematical analysis is carried out is perhaps comparable to the introduction of differential and integral calculus in the seventeenth century. As the Russian mathematician L. V. Kantorovich pointed out, this change was first manifested in a new approach to a variety of problems in mathematical analysis. In this new framework, the investigation of individual functions and the relationships and equations involving them was replaced by the study of families of functions, usually referred to as function spaces, and of transformations and mappings of those spaces by functional operators.

Applied mathematicians and numerical analysts formulate models of natural phenomena or engineering processes by setting up mathematical equations or inequalities to which they then seek solutions, exact or approximate. These may be finite or infinite systems of ordinary or partial differential equations with initial or boundary conditions, integral equations, variational inequalities, or functional expressions with various types of constraints, on which some optimization then has to be done. From the point of view of functional analysis, all such problems can be formulated using operators, linear or nonlinear, that map elements of one well-chosen function space to another. The solutions that one seeks then appear as points in these spaces. Approximation procedures and computational methods that work in this general setting, therefore, are also applicable to large classes of such problems. This becomes an efficient and elegant way to develop effective and reliable computational schemes for approximate solutions of equations arising in the original models.

The origin of functional analysis can be traced to the theory of operators in infinite-dimensional spaces developed by the German mathematician Hilbert, to the general theory of linear normed spaces due to the Hungarian mathematician Frigyes Riesz, and to the Polish mathematician Banach. Interest in this field became very intense as theoretical physicists recognized that the framework of linear operators in Hilbert spaces was crucial for the evolution of quantum mechanics. Efforts by several Soviet mathematicians (Kantorovich, Vainberg, Sobolev, Krasnosel'skii, and many others) added a great impetus to this field. Applied functional analysis began with the publication "Functional Analysis and Applied Mathematics," by Kantorovich, in 1948.[68] In the introduction to this paper, Kantorovich wrote:

> Explicitly, we want to show that the ideas and methods of functional analysis may be used for the construction and analysis of effective practical algorithms for the solution of mathematical problems with just the same success as has attended their use for the theoretical investigation of these problems.

Kantorovich's treatment of Newton's iterative method is an excellent example of the effectiveness of this framework, and also of the power of functional analysis as a tool for use in numerical analysis and computation. Work in this field occupied the attention of several research workers at MRC.

Early work in functional analysis at MRC focused on the solution of nonlinear integral equations. Philip Anselone, a permanent member during the early years of the center, organized an advanced seminar on the topic. This was probably the first meeting on this topic in the United States, and the proceedings launched the first book in English devoted to nonlinear integral equations in their manifold aspects. One of the significant contributions made by Anselone in this area is the notion of collectively compact operators.[69]

Numerical solution of integral equations (both linear and nonlinear) is an important topic in applications, because these equations can model a wide range of problems in science, engineering, and economics. The framework of functional analysis, on the other hand, is quite suitable for analyzing such problems. Typically, these equations are approximated by a sequence of quadrature formulae, and the resulting linear equations are solved by algebraic methods. The logical questions to be answered are then, Do the approximate equations have solutions? If so, do the approximate solutions converge to a solution of the integral equation, and what is the error in the approximation? The theory of collectively compact operators developed by Anselone at MRC[69] provides definitive answers to these questions. To amplify this, we briefly describe the idea of collective compactness and review some of the influence that this work has had.

Most statements about linear operators on finite-dimensional spaces have nice generalizations to a certain class of operators on infinite-dimensional spaces, namely the compact operators. Compactness is a property of certain sets and mappings associated with them. This property is important in numerical analysis for a variety of reasons, some of which are listed here.

- Compact operators have useful properties, which make them easy to analyze and approximate.

- Compactness is a basic tool in establishing the convergence of sequences of approximations because it makes it possible to extract convergent subsequences.
- Compactness is in several ways equivalent to being "nearly finite dimensional."
- A continuous function attains its extreme values (maximum and minimum) on a compact set.

An operator T that maps a space X into a space Y is *compact* if T maps bounded sets in X into precompact sets (that is, sets whose closures are compact) in Y. If Q is a family of mappings $T: X \to Y$, then Q is *collectively compact* if for every bounded subset S of X, the union of images of S in Y under elements of Q is precompact.

Anselone's work influenced many doctoral students at MRC and across the world. For instance, Kendall Atkinson, one of the former students of Ben Noble, has provided a definitive treatment of the numerical solution of integral equations of the second kind. Atkinson combined topics like product integration (from Noble) with Anselone's work. Noble was a major contributor in the field of numerical solution of integral equations. He, along with many of his students, used the notion of prolongation and restriction operators.[70] Numerical solutions of operator equations (such as the integral equations) involving a solution x in an infinite-dimensional space X are usually obtained by finding an approximation x_n in a finite-dimensional space X_n. Prolongation and restriction operators provide a framework for comparing the exact and approximate solutions. A linear operator p_n: $X_n \to X$ is called a *prolongation*, and a linear operator r_n: $X \to X_n$ is called a *restriction*. The operators p_n and r_n are assumed to satisfy certain conditions and relations that are easy to check in specific applications. There are several points of contact and points of differences of this work with related theories developed by Russian and French schools. Using this theory, Noble and his students produced numerical schemes for solutions of linear and nonlinear integral equations, with a nice theory of error in product integration and numerical examples. Many papers and books resulted from this effort.[71]

Variational principles play an important part in mathematics and in physical and engineering sciences because they provide a unified view of different fields, provide a better theoretical understanding, and lead to powerful methods for calculation. The well-known Euler–Lagrange principle can be used to derive basic field equations of many kinds, extremum principles lead to new estimates for important physical quantities, and direct methods form the basis for very accurate computations. Formulation of an equivalent variational problem is not an easy task.

In many problems in applied mathematics, there are two dual (or complementary) variational principles. By choosing suitable trial functions, one can obtain approximations to the physical quantities involved, without having to solve the original equations exactly. For a certain class of problems, one can get upper and lower bounds to a functional that has physical significance, such as energy. In the 19th century, there were isolated attempts in this direction, like the Dirichlet and Thompson principles in electrostatics. Noble made significant contributions to this area, noticing the fact that many basic equations in applied mathematics could be formulated in Hamiltonian form, which originated in analytical dynamics. He obtained effective upper and lower bounds for a variety of boundary value problems. He was also pivotal in influencing the work of several visitors to the center.[72]

Linear equations and generalized inverses

The equation $A\,x = b$, where A is an m-by-n matrix consisting of mn numbers (elements) arranged in a rectangular array of m rows and n columns, x and b are column vectors with n and m elements, respectively, and b is known and x is unknown, represents a set of m linear equations in n unknowns. When all the elements are known exactly, we face three distinct mathematical possibilities: (a) the equations have a unique solution, (b) they have no solution, or (c) they have an infinite number of solutions. Serious complications can arise, however, when elements are not known exactly and/or rounding errors occur. This is usually the situation in practice. The three possibilities can become blurred, as we will discuss later.

If $m = n$, then A is said to be a square matrix. A square matrix is said to be *nonsingular* if its rows are linearly independent. Every nonsingular matrix has a unique inverse A^{-1} such that $AA^{-1} = A^{-1}A = I$, where I is the *identity matrix,* a square matrix of zeros except for units on the principal diagonal. The idea behind the generalized inverse of an *arbitrary* matrix A is to find a matrix associated in some way with A such that: (1) it reduces to the inverse of A if A is square and nonsingular, and (2) it has some of the properties of the usual inverse. Thus, the theory of generalized inverses has its roots essentially in the context of so-called "ill-posed" linear problems. These include problems in which one specifies either too much information ($m > n$) or too little information ($m < n$). These problems cannot be solved in the sense of a solution of a nonsingular problem. For instance, when $m = n$ and A is nonsingular, the above linear equations have a unique solution given by $x = A^{-1} b$. On the other hand, in many cases, solutions of a system of linear equations exist even when the inverse of the matrix defining these equations does not exist. Also, in the case when the equations are inconsistent, one is often interested in some generalized notion of a solution, such as a least-squares solution in which we seek a vector that minimizes the sum of the squares of the residuals.

These problems, along with many others from fields ranging from numerical linear algebra, optimization, and control, to statistics and applied mathematics, can be readily handled with the concept of the generalized inverse (pseudoinverse) of a matrix, and more generally of a linear operator. E. H. Moore in 1920 generalized the notion of the inverse of a matrix to include all matrices, rectangular as well as singular. Moore's definition of a generalized inverse of a matrix A is equivalent to the existence of an n-by-m matrix G such that $AG = P_A$ and $GA = P_G$, where P_X is the orthogonal projector onto the space spanned by the columns of the matrix X. Apparently unaware of Moore's work, R. Penrose showed in 1955 that there exists a unique matrix B satisfying the four relations

$$ABA = A, \quad BAB = B, \quad (AB)^* = AB, \quad \text{and} \quad (BA)^* = BA,$$

where the asterisk (*) denotes the conjugate transpose. These conditions are equivalent to Moore's conditions. The unique matrix B that satisfies these relations is usually referred to as the Moore–Penrose inverse. Penrose also established that this generalized inverse satisfies a least-squares property.

MRC's interest in these matters started very early. Langer had decided that MRC could contribute to the education of mathematicians in Army laboratories by providing what were called Orientation Lectures. These tutorials, intended to illuminate recent relevant developments in mathematics, were presented at the specific request of Army agencies. Noble presented one such set of lectures on Matrix Algebra and Its Applications. Matrices were widely used in applications and essential in computation in the still early days of the computer revolution. One result of these orientation lectures was a request from the Army Map Service for help in solving large systems of linear algebraic equations. By "large" they meant equations with at least 100 unknowns, sometimes even much larger. MRC Technical Report 644 (1966) summarizes progress made on these interactions between MRC and the Army Map Service. The report illustrates the state of the art at that time. An example discussed was a least squares problem involving an inconsistent set of 321 equations in 119 unknowns. The classical method of solving $Ax = b$ by the least-squares method was to solve the system $A^TA\,X = A^Tb$. This produced a set of 119 equations in 119 unknowns. The Army Map Service found that a computer solution of this system by the standard method of Gauss elimination did not produce satisfactory results. On looking at the 119 "pivots" (certain numbers used in the calculation), the reason became clear. The first 118 pivots were of order unity, but the 119th was of order 10^{-13}. This made it clear that the rank of A^TA was almost certainly 118; the 119th pivot would have been zero except for roundoff error in the numerical computation. Essentially, the data set that the Army Map Service was using did not contain enough information to define the 119 unknowns uniquely, and this was the reason why the numerical solution process had run into difficulty

MRC's involvement in this area consisted of much more than simply interacting with practical users. Noble's classic text, *Applied Linear Algebra,* and several papers and books by Thomas Greville are a good measure of activity in this fertile area of applied mathematics. The initial work at MRC was mainly matrix oriented. However, there was a clear transition to linear operators in Hilbert and Banach spaces, in line with MRC's interest in applied functional analysis. This led to a stream of visitors, notable among whom were Adi Ben-Israel and Zuhair Nashed. Ben-Israel's research included coauthored work with Greville,[73] and Nashed's included editing the monumental symposium volume on generalized inverses.[74]

Professor Thomas N. E. Greville

Thomas (Tom) Greville had an unusual career. Born in 1910, he obtained a Ph.D. in the theory of probability in 1933. In the depression years, such a doctorate was more of a liability than an asset. He worked with life insurance companies until he obtained in 1940 a position with the U.S. Bureau of the Census in connection with life tables. He then spent 21 years with the federal government in various departments. In 1962 he accepted a visiting position with MRC, becoming a tenured member of the faculty in 1963 in the School of Business at the University of Wisconsin. On leave from MRC, he served as an advisor to the National Center for Research Statistics from 1973 to 1976. He retired from the center and the university in 1981 and died in Charlottesville, VA in 1998.

Greville had a quiet, pleasant, and unassuming personality and a wide breadth of interests. He published more than 80 papers, of which 21 are reprinted in his selected works (*Selected Papers of T.N.E. Greville,* Charles Babbage Research Center, Manitoba, Canada; ISBN 0-919611-12-5) under six headings: Matrices and Generalized Inverses, Moving Averages, Mortality Tables, Spline Functions, and Parapsychology. Greville's two books were concerned with spline functions and generalized inverses.

Visitors and MRC

MRC provided an excellent environment for the development of genuine applied mathematics, and visitors to the center formed a very important element of this environment. These visiting members constituted a cosmopolitan group that included leading mathematical scientists from the United States and foreign countries with areas of interest contributing to one or more application domains. At MRC they found lively, ongoing discussion about mathematics and its applications to real problems, along with plenty of other interested mathematicians with whom to interact in pursuit of solutions. A high degree of internal interchange between the researchers at MRC, together with the many outside contacts, was invaluable in achieving a truly effective atmosphere of interdisciplinary study. We outline here two examples of technical accomplishments representative of those that we know resulted from the center's environment. There undoubtedly are many more of which we will never know.

When Professor Karl Nickel of the University of Karlsruhe, Germany, visited MRC in the 1960s, one of his interests was in boundary layer theory. Very little was known about the mathematical background of this theory. Some special solutions were known (Blasius, 1908; Falkner-Skan, 1930; Hartree, 1937; Mangler, 1943). From the engineering point of view, some methods had been constructed for finding approximate solutions (power series: Blasius, 1908; Howarth, 1935; Görtler, 1957; the momentum method: Görtler, 1939), but there was neither an existence theorem nor a uniqueness proof. Nothing was known about the well-posedness of the problem, nor about the asymptotic behavior of the solution for vanishing viscosity. In addition, no error bounds for approximate solutions were known.

Since there was so much to be accomplished from the mathematical point of view, Nickel was quite eager to solve some of these problems. He found the newly developed method of differential inequalities to be a powerful tool for this purpose. In 1958 he had been able to prove the first uniqueness theorem, and later was able to treat nearly all of the problems mentioned. However, there were two exceptions. One of these was existence: the Russian mathematician, Olga

Oleinik, gave the first existence theorem in 1962, and Wolfgang Walter devised a second and quite different existence theorem during his stay at MRC in 1970. The second was the problem of error bounds. After 1964 the only unsolved problem from a mathematical standpoint, at least in the simplest case of a two-dimensional steady laminar boundary layer of an incompressible medium, was the problem of computing bounds for error in approximate solutions. Since the boundary layer equation cannot normally be solved explicitly, one uses approximate methods in practice. Thus, one is never sure of the quality of the results obtained. This is true even with the aid of modern fast computers where one sometimes even derives an approximate result with an incorrect sign, if the problem is sufficiently ill conditioned.

From 1955 until the late 1960s, Nickel tried without success to solve the problem of finding error bounds. A part of the difficulty arose from his inability to devote enough time to this problem, because of his other teaching and administrative duties. Once at MRC in 1970, however, for the first time in his professional career, he had enough time to stay with the same issue for a full six months. The time factor along with the presence of other analysts with whom he could interact contributed to his solution of this problem in 1970.

Nickel's experience illustrates why MRC was such an excellent place for thinking men and women to take the time to contemplate various mathematical issues. This, perhaps the most significant value of the center, was well described by Professor Alexander Ostrowski (University of Basel, Switzerland), who visited the center in its earlier years, when he said, "Scientists need a place like this. The urge to get on and earn a living forces many brilliant men to accept academic positions where they cannot devote full-time to research. This is the best way to get scientific substance out of scientists."

A second example of the effect of scientific concentration combined with interaction at MRC is the investigation of finite-difference analogues of the hypergeometric differential equation. This equation, along with its various limiting cases such as the Bessel differential equation, is the most important second-order differential equation in one variable. It is the most general differential equation whose solution can be represented in a power series. This differential equation has three regular singular points. Equations with four regular singular points occasionally arise in applied problems, but not with the frequency with which Legendre functions or Bessel functions surface. These functions are contained in the class of hypergeometric functions. Solutions to an equation with four regular singular points continue to hold interest.

A natural question is how to find the best finite difference analogue of the hypergeometric differential equation. It should be an eigenvalue problem with a solution that can be found explicitly in series form. Gauss showed that the hypergeometric function satisfies a number of second-order difference equations, and for a long time this set of equations seemed to be the right analogue of the hypergeometric differential equation. Among other results of Gauss are a number of very useful continued fractions from the difference equations. These continued fractions are still one of the best ways of computing important functions.

The natural discrete analogue of the Sturm–Liouville second-order differential equation is the three-term recurrence relation satisfied by orthogonal polynomials in one real variable. The 1964 book of F. V. Atkinson[75] has an excellent treatment of this connection, and it is worth pointing out that part of this book was written while Atkinson was visiting MRC.

Professor Richard Askey (University of Wisconsin-Madison and MRC) and his students have done considerable work in developing recurrence relations and analogue difference equations for special functions. In 1942, in a paper on complex potentials, Racah gave a general difference equation different from one that was found earlier by Chebyshev. He also noted an orthogonality relation, but he was not aware that the functions he was considering were essentially polynomials. Jim Wilson, a student of Askey who realized that these polynomials were essentially those treated by Racah, rediscovered this set of orthogonal polynomials. With the existence of several orthogonal polynomials, Wilson and Askey were able to connect the theory. They concluded that the three-term recurrence relation for these polynomials, and other three-term relations that do not necessarily lead to orthogonal polynomials, are the right finite difference analogues of the hypergeometric differential equation. These difference equations may perhaps play roles in discrete problems similar to that played by the hypergeometric equation for continuous problems. There are strong indications from applications such as population genetics that this may be the case.

This work was connected with the visit to MRC of a senior mathematician, Professor George Andrews. Wilson regularly attended the seminar on special functions, and various remarks made by Andrews led Askey and his student to suspect the existence of these polynomials. Andrews and Askey continued their fruitful collaboration, and many suggestions made to Wilson resulted in finding the general class of polynomials.

Professor Wolfgang Wasow in 1976 accurately described the enriching experience that MRC provided for many visitors: "A large part of the progress in the asymptotic theory of ordinary differential equations in the world during the past 20 years first saw the light in the pages of MRC reports by Harris, Hsieh, Langer, O'Malley, Sibuya, Wasow, ..."

In closing this section, we turn from examples of technical research to a discussion of the personal and social environment within which that research occurred. For any organization whose staff consists principally of visitors, as MRC's staff did during most of its existence, this environment is of critical importance, both to prospective visitors' decisions to come and to their productivity after they are there. For this reason, it is very important to note the significant contribution that spouses made to MRC, especially those of the permanent members. Most of MRC's operation occurred during a period when mathematicians were generally male, though there certainly were exceptions, some of whom held appointments at MRC. But most of the visitors and all of the permanent members were men, and many of these had wives and families. Although a visitor might regard a year at MRC as a wonderful release from teaching responsibilities at the home institution, for the visitor's family it was often a major disruption, involving a move—often to a foreign country—along with problems of housing, schools for the children, and all of the other aspects of settling down in an unfamiliar place with no friends to whom one could turn.

In this situation, the wives of the permanent members worked very hard to make things more comfortable for the visitors' families. Particular leadership came from the wives of the directors, especially from Annetta Hamilton Rosser, who was in that position longer than any other. They organized and led the "MRC Wives' Club," which met frequently for many sorts of activities, from gatherings in homes to bird-watching, and which provided all kinds of help to the visitors' families in getting to know Madison, in meeting other people, and generally in surviving an environment which for many must have been very challenging indeed.

Social conditions are different now from what they were in MRC's time, and more support resources are available now than could be found then. But the example of the MRC wives, and the need for the kind of human community that they helped to create, should be important to any organization that relies heavily on short-term visitors.

Interdisciplinary research and team effort

While several instances of interdisciplinary collaboration at the center were mentioned earlier in this chapter and throughout this monograph, other examples of the *esprit de corps* among MRC researchers require discussion. Success in interdisciplinary research is due primarily to close-knit teams whose members are enthusiastic and readily interact with each other and with those of other campus departments. There is much talk in current times about the contribution of multidisciplinary research. Multidisciplinary is not synonymous with interdisciplinary. The effective interdisciplinary study (see below) that was accomplished at MRC cannot by achieved by simply taking an assortment of people with diverse interests and putting them in close proximity to each other for a few years. MRC achieved significant successes by team effort.

When mathematicians deal with problems from physical or engineering sciences, a team effort usually is most effective. Such problems arising from other disciplines seldom fall clearly into a well-defined single mathematical domain. Given the applied problem, there is a real question as to what is the best mathematical technique required for its solution. Generally, a combination of techniques is likely to work satisfactorily. No one mathematician would be capable of evaluating the relative advantages or disadvantages of the various approaches, whereas a group of mathematicians with varied backgrounds such as were available at MRC can rationally debate the pros and cons and recommend the method of solution that provides the required accuracy at acceptable (or preferably lowest) cost. Another advantage of a team effort is that it is highly likely that at least one member of the team will have the needed proficiency in the application domain in question. A good example of this sort is provided by a problem brought to MRC by the Army Map Service.

The mission of the former Army Map Service, subsequently the Defense Mapping Agency, included making maps of portions of the earth's surface. Using photographs from planes and satellites one estimated a large number of distances between sets of points. When the Army Map Service tried to reduce these and indicate where everything is on an absolute scale, they had two or three times as many distance estimates as they had points. This led to a system with two or three times as many equations as unknowns (an *overdetermined* system). Generally, the Army Map Service used a least-squares fit. The Army Map Service normally created maps of the earth, but it was also given the task to re-map the stars in the southern hemisphere. The estimate was that they would have to deal with 800,000 equations in 400,000 unknowns.

MRC thought it might have the necessary competency to help with this task. They assembled a team using some of their core members (Noble, Greville, Rosen, etc.) and invited participation from a faculty member from the Department of Economics who was well versed in handling large data sets (especially partitioning and tearing methods) and visitors from Purdue University Computing Center and the University of Michigan with competence in generalized inverses. A task force from the Army Map Service spent a week at MRC with this group, ensuring that the right set of issues was being addressed. The team was able to provide efficient and robust techniques that not only helped solve the immediate problem at hand but also offered a framework for reliable extensions.

To have a group such as MRC functioning on a continuous basis creates other benefits. The members of the team develop a rapport that makes group actions more effective. They discuss their problems together, they attend seminars and symposia covering a wide range of topics, and they also acquire some understanding of the mathematical specialties of team members. This collaboration moves close to the ideal interdisciplinary mathematician, who has facility in a wide variety of mathematical areas.

Also, a center whose capabilities are known can attract visitors. Though visitors helped greatly with the Army Map Service problem, they came primarily to work with the experts of MRC in their respective fields. A center also provides a useful continuity. The results of the study for the Army Map Service appeared in a MRC technical report (TSR 644, April 1966). More than a year later, a medical problem came to MRC from the Walter Reed Army Hospital, and the solution of this problem was facilitated by one of the techniques that had been studied in the Army Map Service star-mapping query.

Conclusion

MRC provided a rich intellectual environment where resident mathematicians continuously interacted with a stream of visitors both domestic and foreign. It was indeed an intellectual crossroads in the world of applied mathematics, and for a considerable time it was preeminent in the world in that respect. The center provided an excellent environment for the development of genuine applied mathematics, and visitors to the center formed a very important element of that environment.

Many senior members of the center had a rich history of working on applied problems generated either by other disciplines or real-life applications. The work on spline functions is an excellent example of this. The foundational work on splines, especially that of Schoenberg, de Boor, and many of their students, has been very well recognized worldwide. The subject of spline theory and applications has mushroomed, yielding results in many domains ranging from the qualitative theory and numerical treatment of partial differential equations to CAD/CAM, multiresolution analysis of images and patterns, statistical estimation, and regression analysis procedures.

Examples of other significant intellectual contributions of the center include the pioneering work on viscosity solutions. The theory and applications of viscosity solutions is now a very active field with an impact on areas such as optimal control, image analysis, dynamic programming, and game theory.

Functional analysis was one of the most significant mathematical developments of the early twentieth century. The influence of MRC in spreading the use of functional analytical methods in the development of mathematical programming, optimization, variational methods, approximation theory (including spline functions), and general numerical methods was another key contribution of the center.

Chapter 5: Impact: What Some Others Think

Large, vertically integrated research groups which are self-sufficient in terms of mathematical modeling, numerics, and a coherent application focus, are rare. However, science and technology require deliverables which vertical integration can achieve. The MRC provided the glue, the vision, and a measure of leadership for the applied mathematics community to achieve a degree of vertical integration.
—James Glimm, Chair, Department of Applied Mathematics and Statistics, State University of New York, Stony Brook

MRC was the meeting ground for at least two generations of mathematical scientists. Visitors from all over the world passed through this center, enlarging their intellectual horizons. Young and old left their impact and carried with them the rich traditions of the center. MRC had a significant influence on young researchers in shaping their professional careers. As an important input to this monograph, we contacted several visitors to the center for their recollections, experiences, and perspectives about the MRC. The responses were very illuminating. We would have preferred reproducing several of these personal reflections verbatim. However, with one exception, we restrict ourselves to only a few selected excerpts because of space limitations. We quote James Glimm's comments in full at the end of this chapter, as we feel that they describe succinctly and very well the essence of MRC's contributions to the world of applied mathematics.

MRC had a tremendous influence on shaping the professional direction of several young scientists. Visitors from Asia and Europe found this place as rejuvenating as those from the American continent. Klaus Ritter (professor, Technical University of Munich, Germany), then a young postdoctoral visitor, has this to say: "In my opinion one of the most significant contributions of MRC was to provide excellent opportunities to young postdoctoral researchers from all over the world to come to Madison and work in an environment that was very conducive to research. After receiving my doctoral degree in mathematics in 1964 from the University in Freiburg I had the privilege to spend two years (September 1965 through September 1967) as visiting assistant professor at MRC. This visit had a profound and lasting influence on my life. During those two years I was exposed to numerous new and interesting aspects of mathematics and I had an opportunity to meet many stimulating people. I am certain that without this experience I would not have pursued an academic career." He concludes his observation with a profound thought that "I had hoped that some of my doctoral students would also have an opportunity to spend some time at MRC. All I can do now is to try to convey to my students some of the excitement I felt in those days. In a recent course evaluation by my students there were remarks to the effect that they find these reminiscences very motivating. A final tribute to MRC."

Another telling and contemporary observation is from Joseph W. Jerome (professor, Department of Mathematics, Northwestern University), who completed his dissertation in 1966 under Professor Michael Golomb. The latter was a frequent visitor to MRC. It was but natural for Jerome to spend a two-year postdoctoral stay at thRC. In his own words, "When I arrived, I found a vigorous research center, active in most areas of constructive, applied, and computational mathematics. It was a center which was comprised of a core group of senior mathematicians, including J. Barkley Rosser, the director, and several others, notably Phil Anselone, Tom Greville, Ben Noble, and Iso Schoenberg, as well as a considerable group of visitors, both junior and senior." And he continues, "For me at that time, the presence of Iso Schoenberg was invaluable. His mastery of constructive analysis, particularly as applied to spline theory, was extremely informative. The expositions of Tom Greville were illuminating. The concrete uses of functional analysis by Phil Anselone and Ben Noble helped to solidify my own adaptation of such an approach, begun under Golomb's direction. I also availed myself of several seminars in the mathematics department."

Breadth of expertise and evolution of areas of major significance made MRC a unique place in the world. Those who had the good fortune of association with the center derived tremendous benefits in their professional careers. Ram P. Srivastav (professor, Department of Applied Mathematics and Statistics, State University of New York at Stony Brook) states:

> I knew very little about MRC before I went there. Rudolph Langer must have been a seer and also a very resourceful person. I was familiar with the symposia organized by learned societies at their annual meetings. The attempt to define the state of art in an environment of expanding knowledge is very commendable. My guru, Ian Sneddon, had contributed a paper entitled "Boundary Value Problems in Thermoelasticity" in the 1959 symposium, Boundary Problems in Differential Equations, but what a galaxy of people were assembled: Bellman, Birkhoff, Collatz, Courant, Diaz, Fichera, Friedrichs, Henrici, Varga, and Young. We get to learn whatever is of immediate research interest anyway, but these symposia and subsequent publication of proceedings were a valuable contribution to knowledge. A number of people who went through MRC developed into visible scientists, Askey, Conley, de Boor, Rabinowitz, Parter, Mangasarian come to mind quickly and the list I am sure is very long. Askey's results on positivity of certain partial sums played a crucial role in de Branges's proof of Bieberbach conjecture. The persons who had most influence on my future work were Schoenberg and de Boor, though we never worked together. MRC had a truly international character. Where else could you see Ben-Israel and Nashed together! There was a contingent from England and I recall Aubin being there too.

MRC had a major impact on the direction of applicable mathematics, both nationally and on the global scene. Dr. John V. Ryff, a program director (retired) at the National Science Foundation, has this to say: "It goes without saying that MRC set the standard for mathematical research in areas such as applied analysis and in recognizing computational mathematics long before the establishment understood the need for such investments. Other institutes do not only emulate the MRC model, but we also see the focus of entire departments shifting to areas long emphasized by ARO and your colleagues. At least that is what I see in *SIAM News*, which itself shows dramatic shifts almost before your eyes. I also think those acquainted with the MRC mentality were able to shift more easily as trends change." Mathematicians and users of mathematics from government and industry also benefited immensely from the center's products and activities (publications, conferences, symposia, tutorials, and contacts). In the words of Dr. Gerald R. Andersen (research mathematician with various Army organizations, now retired), "My involvement with the MRC during this period (1972–1975) was solely as a mathematician. I had occasion to read a number of the MRC technical reports and in later years used some of their results in applied problems. I made one trip to the University of Wisconsin in this time period to attend an MRC symposium on advanced spectral analysis. It was outstanding. I had a deep respect for the MRC mathematicians and the work they accomplished for Army scientists and mathematicians under the MRC contract. Of particular importance was the work of Carl de Boor on B-splines and their application to various areas of numerical analysis at Army laboratories. In the area of statistics, the work of George Box and his British colleague Gwilym Jenkins gave all time series workers a new and important way of modeling stochastic time series that did not require frequency domain information. Thus statisticians everywhere were freed from having to compute and interpret the spectral density of a time series in order to forecast the future behavior of a series. This had the consequence that their methods allowed nonmathematicians and nonscientists to produce quantified estimates of time series models and to forecast time series with analytic characterizations of the forecast errors."

How did MRC influence other comparable activities in the mathematical sciences? What were its relations to the traditional departments (mathematics, statistics, and computer sciences) on the campus? We will explore these issues in some detail in our concluding chapter. However, we would like to quote here some relevant comments on this topic. Once again, Jerome observes, "Younger mathematicians may see a clear parallel between the IMA and MRC, but I do not. MRC was less interdisciplinary, but the era did not yet encourage such close cooperation. MRC was not

Bell Labs. It did, however, foster intense investigation surrounding the development of 'useful and concrete' mathematics. It also fostered the careers of mathematicians at every level by an emphasis upon scholarly activity. I owe a large measure of my career to MRC, and I shall always be grateful."

A thought-provoking observation comes from Kendall E. Atkinson (professor, Department of Mathematics and Computer Science, University of Iowa): "Regarding the MRC and the IMA at Minnesota, I believe that Madison eventually lost a freshness in what it was doing. Friedman, Miller, and others at Minnesota created a new and vibrant way of focusing on new research topics in applied mathematics. The new research was increasingly applied, interdisciplinary, and group oriented. The IMA created a new kind of structure, and it seems well suited to the kind of research now being done in the USA. The Madison MRC's framework did not change sufficiently. I do not know if it could have been fixed at Madison. The simplest approach was simply to start over again, as was done at the IMA. As regards the center at Cornell University, I have had no contact with it. I believe it has been important, but more for Cornell than elsewhere.

"For its early years, I do not believe that many people in the Mathematics Department at Wisconsin had a high regard for the MRC. This was their error, and most people eventually realized such. I was only dimly aware of the politics and related matters involving the MRC and the associated academic departments. I learned more about such matters when I began my first real academic job at Indiana University in the fall semester of 1966. The book *The Masters* by C. P. Snow has much to recommend it as an introduction to academic politics."

We conclude this chapter with the reproduction (unedited) of comments by Professor James Glimm (chair, Department of Applied Mathematics and Statistics, State University of New York at Stony Brook) on MRC and its successor institute, Cornell's MRI, as well as the Army High Performance Computing Research Center (AHPCRC) at the University of Minnesota:

> The single most important variable in determining the success of a Center is the quality of its leadership, especially of its director.
>
> MRC, during the years I interacted with it, had an energetic and skilled leader, John Nohel, who more than any other single force, determined its success. Immediately following in importance are the principal supporting personnel, the lead faculty, and sponsor organization. This leadership team for MRC defined a mission, which made sense from the multiple points of view of the different major groups that MRC served. These included: Army labs and the Army research needs, the applied and numerical mathematics community, and the University of Wisconsin and its mathematics department.
>
> These were tensions in keeping a common focus among these groups, some highly politicized. During John Nohel's tenure as Director, these tensions were largely under control.
>
> MRC functioned through supported post docs, graduate students, and academic visitors. Its impact was multiplied through its periodic conferences.
>
> During the years of John Nohel's tenure as Director, during the Cornell MRC follow on, and to this day, applied mathematics is to a large extent a cottage industry. Large, vertically integrated research groups, which are self-sufficient in terms of mathematical modeling, numerics, and coherent application focus, are rare. However, science and technology require deliverables which vertical integration can achieve. The MRC provided the glue, the vision, and a measure of leadership for the applied mathematics community to achieve a degree of vertical integration. In the process it served the Army research needs, as these became foci for this integrated leadership.
>
> It is my feeling that John Nohel was greatly aided in this integration process by the tireless energy and insightful judgements of Jagdish Chandra.
>
> In some ways, MRC marked the high points and the end of an era. This was the era of theoretical, analytic, and asymptotic methods. These methods did not go away, but

they lost primacy to numerical methods. MRC did not make this change, one of two facts, which could have been the cause of its termination. The second factor was old wounds from past political battles, which deprived the university of the competitive will to campaign effectively for renewal of MRC.

Why did MRC not make the change to numerical dominated methods? Because its senior faculty were too deeply enmeshed in the older methods. It is a specific example of technological renewal and innovation outpacing the humans who conduct it. This is an issue whose importance has only increased since those days.

Why did Wisconsin not mount an effective renewal bid? Again, in my opinion, there was a failure to grasp a fundamental change. Wisconsin no longer "owned" MRC and had to compete against very serious contenders. Old battles clouded realization of the value of MRC to Wisconsin until it was too late.

So the effort moved to MRI at Cornell. I have observed that the skill needed to win a Center award can be quite different from the skill need to manage and lead it.

As with MRC, the strength of MRI was a type of integration of applications, physical modeling, numerics, and applied mathematics, which is difficult for a single investigator to achieve. The existence of such a center, with its outreach programs, however, allows single investigators to benefit from and participate in a vertically integrated research activity.

MRI also had its problems, including faculty in some areas who were not committed to this vision of applied science. The evolution in technology and its management continued, and tensions over deliverables vs. basic science became deeper. These changes occurred at a national level and filtered down to MRI. Probably the two factors which spelled the end of MRC were also undoing the MRI, and a transfer of the Army effort to Minnesota's AHPCRC.

During Nohel's period as director, MRC was the national leader (together with the Courant Institute) in defining the main ideas and research themes for applied mathematics. These were productive years, in which applied mathematics made great strides. The credit for these accomplishments can be shared among Nohel, Chandra, MRC, and ARO, as well as the senior MRC faculty.

Chapter 6: MRC: Some Lessons Learned

There are good reasons for studying history, but it can also trip one up. On the one hand, Santayana argues that those who ignore history will be condemned to repeat it. On another, the philosophic pitcher Satchel Paige advises, 'Don't look back; someone may be gaining on you.' I don't go completely for either one.
—Vannevar Bush, in *Pieces of the Action,* 1970

In the post–World War II era, MRC was one of the earliest comprehensive examples of collaboration between the government and a university. By taking a broad view of mathematics that embraced pure and applied mathematics, the center provided a model of an interdisciplinary effort that interacted very well with the spectrum of sciences. Traditionally, mathematics and the sciences have grown together via mutually beneficial interactions. MRC was a good example of such a relationship. This successful model of synergy has been emulated and adapted in various contemporary activities. While the MRC paradigm underscored the promise of interdisciplinary activities, the center's history also highlighted key issues in implementing this model.

Key strengths of MRC, and some of the lessons learned

This section first summarizes some key strengths of MRC that previous chapters have illustrated, then discusses some of the lessons learned. The following lists some of the strengths.

- Relatively stable funding from the U.S. Army enabled the center to attract several outstanding mathematical scientists as the core of MRC.
- The structure of MRC fostered, prolonged, and sustained dialogue between the core members and visitors to the center.
- Physical proximity over a sufficient length of time resulted in substantial contributions via joint publications, technical monographs, and the development of new courses.
- The center maintained a mix of senior and junior researchers. This setting was a rich environment for training students, mentoring junior researchers such as faculty and postdoctoral students, and promoting meaningful exchanges among senior researchers.
- The research efforts at the center spanned a broad range of mathematics and its applications, which evolved over a period of time. MRC was a major source for the creation of a genuine applied mathematics program that connected advances in mathematics with well-developed application domains.

A major premise of the MRC model was that research is a collective activity. Even the proverbial "loners" benefit from the stimulation generated around them by other interacting groups. Many resident members and visitors to the MRC would testify to such a positive influence. The richness of an active research environment, the new ideas and innovations, and the inspiration generated by such a milieu benefited all those who participated. To a large extent, the permanent (core) members at the center played a key role in generating such a stimulating setting. After all, it is rarely wise to entrust a single junior-level faculty member or researcher with either the launching or the maintenance of a major research activity.

MRC, as it was originally conceived, was to have a limited permanent staff of highly competent and versatile mathematicians who would provide leadership and assure continuity of effort. The acquisition of such a staff presented tremendous challenges especially to the early directors of MRC. This problem was further accentuated in the early years, due to unprecedented transformation of the working conditions and opportunities for the country's research mathematicians. However, when the center eventually was able to hire permanent members, it offered them life appointments instead of term appointments (e.g., five years subject to renewal). These appointments were always made jointly with one or more of the academic departments of the university. This became a serious matter in later years, as the experience and level of research activities of some of the permanent members were not appropriate for the center's evolution towards new technical directions.

For instance, the MRC realized quite early the important role of scientific computing in the development of applied mathematics. Partly due to its structure, the center could never make a successful movement toward that goal. One of the hard lessons gleaned from the MRC experience is the need to ensure enough flexibility in staffing structure for any long-term and major interdisciplinary research activity.

The academic department is the unit within which university research activity traditionally takes place. Although interdisciplinary research often involves several departments, faculty and their students ultimately perform it. Therefore, its success depends on the overall climate. MRC had many attributes of a successful department, including research leadership and scholarly activities such as the tradition of mentoring younger scholars. Mentoring is often a powerful force in these academic activities. However, such traditions do not simply occur but rather must be developed. A positive research ambience includes a range of features. Most research is a collegial activity, and collaborations can involve either research on common interests or research on related topics. In this regard, a simple matter like a common area where research can be discussed in a collegial manner becomes crucial. Periodic research seminars in which faculty and students can regularly participate are equally important. These interactions among faculty, visitors, and students generally result in new ideas, new formulations, and possible answers. MRC had a rich tradition in all of these aspects and can boast an excellent track record.

Although MRC was not a formal teaching department, it had a significant impact on the educational programs at the University of Wisconsin-Madison, especially in the mathematics, statistics, and computer sciences departments. Its involvement offered two contributions. First, MRC was a catalyst for training a new generation of mathematical scientists in cross-disciplinary thinking. Second, the center stimulated researchers and faculty to expand their intellectual horizons. The center also provided several case studies of strong and influential leadership in applied mathematics. The brief profiles of four directors of the center highlighted in Chapters 1–3 offer confirmation of this statement.

MRC: A paradigm for center-based research

One definition of *paradigm* is a pattern or example. We consider MRC as a paradigm of center-based applied mathematics in a university environment, as contrasted to the traditional department-based paradigm for conducting such research. The latter often operates through single investigators who obtain personal grants from funding agencies and can then use these to support graduate students and perhaps junior faculty in the department.

A unifying force and catalyst for MRC was its broad definition of applied mathematics and its relevance to properly chosen application domains. As an alternative to the department-based model that has flourished over several decades, the center-based research model focuses on common themes, sets of problems, application domains, or methodologies. Such centers are often referred to as institutes, laboratories, or programs, and in the following discussion we employ the term "center" generically.

Centers relate to academic departments in several ways. They can be included in an academic department, and annual block grants of a few hundred thousand dollars can sustain such an effort. Centers may also draw faculty from multiple departments at the university. MRC operated in the latter mode. Most of its core staff had joint appointments with the Departments of Mathematics, Statistics, or Computer Sciences. As MRC's research agenda evolved, faculty members from several engineering departments were drawn in. The steady funding streams that MRC enjoyed over most of its life, as well as a cohesive administrative structure, allowed for stability in such a configuration.

In some respects MRC was a *freestanding* center, in that it was independent of departments. Such a center can draw faculty members from several departments to serve in a part-time or full-time capacity. Such centers tend to be treated as independent entities, and their strength depends heavily on the quality of the leader or director.

A freestanding center approaches a department in its autonomy, but it differs from a typical academic department in several key ways. For example, its ability to grant degrees or to offer tenure to faculty is limited or nonexistent. However, a major advantage of a freestanding center is that it is not constrained by the norms of established disciplines or subdisciplines, and it is therefore more likely to cross disciplinary barriers and to conduct interdisciplinary research. MRC had some success with this mode of operation. Unfortunately, there were several instances when its role as an independent center resulted in considerable strain between the center and the departments closest to it (mathematics, statistics, and computer sciences).

Other advantages of center-based research include the ability to permit and encourage bold and progressive research. Moreover, this type of research often supports fresh ideas, new approaches and methodologies, and diverse sets of application domains. Discipline-based research, on the other hand, sometimes tends to be restrictive in addressing emerging issues in engineering, science, and technology. When centers are staffed from several departments, they have the potential to increase university collegiality and enhance intellectual dialogue. In addition, such a model reduces parochialism by supporting cross-pollination of ideas.

While center-based research yields significant gains, it also has significant disadvantages that can weaken the ability of the model to function. First, it can be an intellectually risky proposition. The traditional department-based research has well-established disciplinary metrics and standards against which progress can readily be evaluated. Center-based research, on the other hand, does not always lead to uniformly accepted measures of success. This issue has created considerable difficulty among academic groups, particularly during promotion and tenure processes. In order for such interdisciplinary activities to be fully acceptable in a university setting, it is essential that the reward and retention procedures be appropriately adapted to their circumstances.

The center-based model also has the potential of instability. One of the greatest advantages of MRC was its stable and predictable funding from the U.S. Army. Basic research requires long-term fiscal projection. Otherwise, the quality of both the work and the workforce can suffer deleterious effects, especially if the center has to depend on highly fluctuating grants and contracts. Moreover, serious problems can occur if short-term or ill-conceived priorities of sponsoring agencies weigh heavily on a center's work. In this regard, the role of the center director and senior staff can be very crucial. MRC was fortunate to have some outstanding and articulate technical leadership during its existence.

An organizational concern also arises with the center-based research model in that it can create rivalry among the faculty and actually widen the chasms among departments. Those who are engaged in center activities might benefit from reduced teaching loads, greater research support, better space and facilities, and monetary rewards in the form of higher salaries. Those in the traditional setting of an academic department may resent this apparent disparity. The accounts offered by four professors in the Department of Mathematics at the University of Wisconsin during the early years, in which departmental phone lines were shared while MRC was generously funded to afford a private line for each visitor, clearly demonstrate this conflict.

The limitations of the center-based model as highlighted by the MRC experience indicate the need for balance between this paradigm and the traditional approach in order to advance the research agenda. One essential lesson from MRC's history is that the various models of operation should not only coexist but also work toward a complementary relationship. Not all research is best suited to the context of a center or interdisciplinary environment. Department- and disciplinary-based research is both valuable and essential for intellectual creativity. However, there are areas where boundaries need to be broken, and here the center-based model has an advantage. Investment in research should aim for a balanced portfolio, unhindered by traditional boundaries but using disciplinary success as the foundation.

A comparison with four other research centers

In this section we compare the method of operation at MRC with those of four other research centers: the Courant Institute of Mathematical Sciences at New York University, the Mathematical Sciences Research Institute (MSRI) at the University of California, Berkeley, the Institute for Mathematics and Its Applications (IMA) at the University of Minnesota, and the Institute of Pure and Applied Mathematics (IPAM) at the University of California, Los Angeles. As we will see, there are similarities among the five centers, but also significant differences, some of which (in the later centers) probably arose from observation of the MRC experiment.

Before MRC was founded in 1956, research in applied mathematics was mostly a cottage industry. Universities were small, and members of a mathematics department were expected to share teaching and administrative chores. Those who wished to do research in pure or applied mathematics did so on their own time, in addition to other duties. Faculty members active in research were sometimes given relief from administrative duties. Apart from that there was little independent research support available.

An important early exception in the United States is the Courant Institute at New York University, which started off small in the late 1930s. Over the past fifty years, this institute has been a leader in the promotion and integration of carefully chosen areas in the mathematical and computer sciences as a unified field.

A brief history can be found at the Courant Web site (www.cims.nyu.edu). It is interesting that Richard Courant came to NYU in 1934 as a visitor, leaving his position as the director of the Mathematics Institute at the University of Göttingen in Germany. Courant, together with K. O. Friedrichs, J. J. Stoker, and others, formed a cohesive group. During World War II, supported by OSRD, they worked on various war projects. After the war, ONR and other government agencies supported and encouraged the group. The Atomic Energy Commission installed a state-of-the-art electronic computer at NYU in 1952. Groups were formed to study electromagnetic wave propagation and plasma fusion. It was at this period, around 1960, that the group acquired a large measure of autonomy within the university. The computer science program began in the 1960s and ultimately became a separate department, with close contacts with the Courant Institute to this day. In 1978 the institute formed a group on computational fluid dynamics. More information about the remarkable diversity of the institute's research activity and funding at the present time can be found at its Web site (www.cims.nyu.edu).

Courant and MRC both benefited from large numbers of visitors and stimulating programs of research seminars. These seminars promoted interactions among faculty, visitors, and students, thus paving the way for fruitful collaboration. This is one of the best by-products of any properly conceived interdisciplinary activity. For many years, MRC and Courant provided excellent examples of how such programs work. A key feature of the two institutions has been the inclusion of senior and junior researchers in a productive environment for an extended period of time. Later centers have replicated this mode of operation.

The Courant Institute had, and still has, clearly defined subareas, with the emphasis changing with time. Conversely, MRC had a more diffuse structure. In MRC there were subareas, such as statistics, of direct and immediate importance to the Army. There were other subareas, depending on the interests and background of the director, the interests of the permanent members, and those of the large number of visitors, that might have less immediately visible connections to Army interests. Visitors were selected according to merit and to their potential for contributions to subareas that, with some exceptions like spline functions, were often not precisely defined.

A natural question arises, Why does the Courant Institute still exist at the present day, whereas MRC folded in 1987? Here are some reasons, based of course on history and hindsight.

- A cornerstone in the development of MRC was that of *core competency,* continuing over long periods of time, in selected areas of mathematics of broad interest to the Army. MRC tried to achieve this through the appointment of permanent members in some of these key areas. Originally this was a strong point of MRC, but it became a source of inflexibility in later years, as we have shown in Chapter 3. The Courant Institute and other later centers had more flexible structures.

- MRC's difficulties started around 1970, at the time of the student troubles connected with the Vietnam War. MRC lost valuable permanent members, had its director forcibly retired by the university administration, and was moved physically to the edge of campus so that contact with departments became more difficult. Relations with some University of Wisconsin-Madison faculty became soured in the climate of that time, and some permanent members of MRC may have been distracted from their concerns with the center's research direction. Centers started since have not had this traumatic experience. In fact, in the times following the demise of MRC, universities have actively encouraged Department of Defense funding in the mathematical sciences.
- In real terms the MRC budget decreased (as did other research funding in mathematics) in the 1970s, while demands on its resources were increasing. When Rosser retired in 1973, the hope was that his successor would have comparable stature and could lead MRC into the future. A number of national figures were unsuccessfully approached, and in the end a director had to be appointed from within the Madison campus. Buck accepted as a transitional director, in order perhaps to allow time for things to settle down. Chapter 3 of this book covered this part of the history of MRC.
- A critical turning point occurred when the Army decided to have independent periodic peer reviews of MRC, in addition to the semiannual review by the AMSC. This was a sign of the times, when research funding agencies demanded visible value for money under different and perhaps shorter-term criteria.
- The results of the first such review in May 1977 are summarized in Chandra's letter to Noble, summarized in Chapter 3. While some aspects of this letter were favorable, it raised key issues: more emphasis on physical mathematics and computational methods, the permanent member problem, and the location of MRC. MRC made attempts to solve these issues, but in the end it did not succeed in doing so. A number of factors contributed to this, some of which have been mentioned elsewhere.

We next consider the two mathematical institutes established by the NSF in 1982: MSRI at the University of California, Berkeley, and IMA at the University of Minnesota.

MSRI hosts between two and four major research programs, which cover a wide variety of topics in pure and applied mathematics. The primary mechanism of financial support is to partially sponsor visiting faculty and to host a good number of postdoctoral fellows. Unlike MRC, there are no core permanent members. The institute has been quite successful in organizing introductory and specialized workshops. In addition, it has initiated new areas of research, often in conjunction with other disciplines (e.g., biology, physics, economics). MSRI has also sponsored activities designed to develop human resources and to increase public awareness and appreciation of research in the mathematical sciences. Further information can be found on the MSRI Web site (www.msri.org).

The IMA mission is more closely aligned with that of MRC in that it aims to close the gap between theory and application. IMA activities allow mathematicians and other scientists to share a stimulating research environment, much as did the activities at MRC. IMA extended the paradigm of contact between researchers of varied backgrounds and experience. Annual programs intended to encourage interaction between mathematicians and scientists from industry, government, and academia are reminiscent of the "years of concentration" at MRC (see Chapter 3). However, one major difference between the two operations is that IMA has exclusively interacted with visitors and postdoctoral fellows, and it thus has no core institutional legacy. In addition, IMA has no requirement to produce hard deliverables for, nor to render assistance to, its sponsoring agency. In the case of MRC, there were specific expectations that the expertise at the center would be available to the U.S. Army. IMA maintains a Web site (www.ima.umn.edu), where further information can be found. A list of 124 books published by Springer-Verlag as part of an IMA Volumes series from 1986 to the present is given at www.ima.umn.edu/volumes.html. These cover a diverse range of topics in the mathematical sciences.

NSF held a competition in 1997–1998 that led to the renewal of MSRI and IMA and the establishment of a third center, IPAM, based at the University of California, Los Angeles. We quote from their Web site (www.ipam.ucla.edu):

> The overall mission of IPAM is to make connections between a broad spectrum of mathematicians and scientists, to better inform mathematicians and scientists about interdisciplinary problems, to broaden the range of applications in which mathematics is used. ... The ultimate goal of IPAM is to bring the full range of mathematical techniques to bear on the great scientific challenges of our time, and to stimulate exciting new mathematics via new problems motivated by other sciences, and to train the people who will do this.

Among the pages on the IPAM Web site, an especially interesting source of further information is www.ipam.ucla.edu/publications/inauguration/.

Interdisciplinary research

When viewed abstractly, interdisciplinary research is a good idea. In fact, combining expertise from several academic disciplines has allowed for many exciting research advances. Even in a broad discipline like mathematical sciences, many scientific advances have occurred at the interfaces of various subdisciplines. However, the promotion of interdisciplinary research in an academic environment has always been a major challenge. As the experience at MRC has clearly shown, such interdisciplinary research crosses different intellectual cultures, including science and engineering. For an individual faculty member to succeed in this environment, he or she must learn the languages and cultures of multiple disciplines. On the other hand, well-structured and balanced teams can offer a necessary framework and facilitate fruitful interaction.

While MRC was not completely successful in this regard, it made significant inroads in interdisciplinary collaboration. A significant part of the incentive for this center was its continued exposure to multidisciplinary problems highlighted by Army scientists and engineers. As von Neumann noted, the most vitally characteristic fact about mathematics is its peculiar relationship to the natural sciences. Traditionally, mathematics and the sciences have grown together with mutually beneficial interactions. The growing numbers of multidisciplinary activities that we see today are building on the initial successes achieved at MRC towards this goal.

The Division of Mathematical Sciences (DMS) of NSF has established a competition for funding innovative proposals for interdisciplinary research in the mathematical sciences that can be accessed through the Web at www.nsf.gov/mps/dms/start.htm. We quote from the NSF Program Solicitation for Mathematical Sciences Research Institutes (NSF-86) for the year 2000:

> The existing institutes [MSRI, IMA, IPAM] meet only part of the increased challenges due to the growing interface of the mathematical sciences with other disciplines and the increasing fundamental and mathematical and statistical problems whose solution will contribute to both the knowledge base and societal needs. Thus, the establishment of new institutes will aid in helping the mathematical sciences, in partnership with science and engineering, to meet these new challenges.

It goes on: "Mathematical sciences research institutes exemplify larger-scale projects that are effective in important ways, . . ." of which some are quoted briefly. "It is expected that additional institutes will demonstrate new, imaginative and different missions and formats from the existing institutes [MSRI, IMA, IPAM]." The Program Solicitation goes on to list five examples "to stimulate thinking." It is anticipated that, based upon availability of funds, there may be up to four new institutes, each to last five years. The Program Solicitation makes instructive reading in connection with interdisciplinary research.

The U.S. Army, MRC, and the mathematical sciences

In evaluating the value of MRC to the U.S. Army, one should take into account the state of applied mathematics in 1956 when the center was established. Analytical methods were predominant, and digital computing was still in its infancy. As described in the section on "The Army and Mathematics" in Chapter 1, the Army showed remarkable prescience in singling out four areas for MRC's mission: numerical analysis, statistics and the theory of probability, applied mathematics, and operations research. The mathematics department at Wisconsin was not very strong in some of these areas at that time. It would be interesting to know why Wisconsin was chosen over more than twenty formal proposals submitted by other universities. But one thing is certain: MRC had the significant effect of strengthening the university effort in these areas and in promoting such research nationally.

As a result of experience in World War II and the years following, various groups in the Army decided that research in the above four areas should be encouraged. It was decided that a research center in applied mathematics should be of value to the Army and the nation. They took a very enlightened view of the word "value." Thus MRC resulted, with fairly flexible terms of reference:

- The objective in establishing MRC was to provide a nucleus of highly qualified applied mathematicians who would conduct research in mathematics that was of interest to the Army, as defined broadly in the four areas above. These were to be supplemented by visiting members, both senior and junior.
- In addition to research, the center could offer essential guidance in bolstering applied mathematics groups in the Army Research and Development Laboratories, and also offer technical assistance to individual Army scientists and engineers.
- Beyond its immediate research and the responsibilities towards the Army, MRC was considered a national resource that would contribute to research in applied mathematics in a variety of ways. For instance, technical reports and publications in refereed journals made MRC research results generally and openly available. In addition, there were twice-yearly meetings with published volumes on recent advances, symposia on topics of current interest, and books published by members, stimulated by their time at MRC.
- Visitors were exposed to a variety of applied problems, and many of these then participated in defense-related projects. MRC increased substantially the supply of mathematicians whose training included exposure to applications.

The Army has pressing mathematical problems that need short-term remedies. Clearly, MRC was not the appropriate place for the Army to take its urgent troubles for a quick solution. An institution like MRC has the capacity to identify broad classes of problems whose solutions help in the resolution of many existing unsolved quandaries. The time scale for such activity tends to be intermediate to long term. The gap between what could realistically be expected from such centers and the expectations of some Army groups led, on occasion, to less than satisfactory relations between the two.

Another serious drawback in this relationship was the lack of qualified scientists and engineers in the Army who could benefit from the broad research accomplished at the center. MRC was adept at developing new knowledge and at technology generation, but technology transfer required additional mechanisms. In order for technology transfer to be successful, there is a need for strong time and resource commitment on both sides of the fence. Based on the experience of MRC, the Army was able to develop more effective methods for technology transfer. This was especially important in the establishment of the Army High Performance Computing Research Center (AHPCRC) at the University of Minnesota in 1989, to be discussed later.

During the 1980s the nature of applied mathematics changed dramatically. Computers became easier to use, and the size and complexity of the problems they could handle was increasing. Many more engineers and scientists were familiar with mathematics at a quite sophisticated level, as needed in their particular subject. Mathematics and computing became important in a wide variety of new areas. The Army response to these challenges evolved over the years. For instance, the U.S. Army Science and Technology Master Plan (ASTMP) for 1998, in its chapter on basic research and referring to mathematical sciences, states

> Mathematics plays an essential role in modeling, analysis, and control of complex phenomena and systems of critical interest to the Army. Mathematical modeling is increasingly being identified as critical for progress in many areas of Army interest. The mathematical scientific tasks in these areas of interest are frequently of significant complexity. As a result, researchers from two or more areas of mathematics must often collaborate together and with experts from other areas of science and engineering to achieve Army goals.

This emphasis on multidisciplinary research has grown over the years. The Department of Defense, in particular, has spearheaded this program under the Multi-disciplinary University Research Initiative (MURI) program, which is now quite extensive. NSF and other federal funding agencies are adopting this mode of support for basic research.

The Army was the pioneer in establishing a multi-investigator research center like MRC; we would now term this a center of excellence. The lessons learned through that experience have been quite valuable to the Army as well as to other federal agencies. For instance, the Army established a Center of Excellence in high performance computing in the form of AHPCRC, established in 1989 at the University of Minnesota. It focuses on the development of new computational methods and techniques that exploit the full potential of extant computing platforms and on the exploration of new computational architectures to solve challenging problems from a variety of fields that are of interest to the Army and the Department of Defense. AHPCRC also has an extensive program of training and education, particularly for the Army scientists and engineers. Things have come a long way from the days of MRC!

In conclusion: Four issues raised in the introduction

The following four issues have appeared throughout our monograph. Here we repeat them and summarize some lessons learned that we have illustrated at length earlier.

- *The difficult tradeoff between appointing key scientific staff for long periods (or even permanently) to provide long-term stability and continuity, and retaining program flexibility by appointing key staff for shorter periods.*

In the climate of the times when MRC was established in 1956 through the 1960s, research mathematicians had job security only if they were tenured members of a university department. Applied mathematicians of the caliber needed at MRC would have tenured appointments in other universities. It was not an attractive proposition to offer them an appointment to this new research center funded by the Army without a guarantee of continued funding. Only a university department could grant tenure. To attract someone as a permanent member of MRC, the earlier directors of MRC (Langer and Rosser) had to approach departments to give tenure to persons they wanted, which meant that these individuals would have to fit into the departments' needs. These directors dealt with that problem by creating permanent members, funded by MRC for as long as Army funding continued. Of course, this term was never clearly defined. In retrospect, it appears that this decision created severe difficulties for MRC. This problem has been resolved in various other ways by research centers established after MRC, some of which have already been discussed.

A reasonable procedure seems to be to appoint a director for five years and to make research appointments for terms of at most five years. In both cases renewal should depend on the circumstances of individual cases. This is certainly possible (and desirable) in the climate of the present day, when mathematical scientists are in great demand, tenure is somewhat in the background, and research opportunities are plentiful.

- *The question of how much power should reside in the director of such a center, and how much should be held by the community of principal research staff, the administration of the host organization, and the sponsoring agency.*

At MRC the power of the first two directors was different from that of the last two (or of Buck during his term as acting director). Langer ran MRC single-handedly, a management style that was in line with the times. Rosser arrived when mathematical sciences and computational mathematics were expanding, and a firm hand was needed to decide priorities. He supplied that firmness, and again this was appropriate for the times.

Times had changed, however, by the time Noble took over in 1975, succeeded by Nohel: they were more limited in what they could and could not do, for several reasons. Delay in appointing a successor to Rosser had left a vacuum. For the first time MRC had a committee of the "principal research staff," and there was also a campus advisory committee on MRC. Because of decreased federal funds, the Mathematics Department took a proprietary interest in MRC and looked to MRC for research support, perhaps not unmindful of the fact that both Noble and Nohel were tenured members of that department. Ultimately the power of the director to implement changes was much restricted.

- *The extent to which the culture of the host organization, its rules, personnel practices, and departmental policies can restrict the freedom of action—and even the survival—of an independent center.*

In MRC's case the "host organization" was a large state university with 40,000 students, with a strong leftist element on the campus. Below the chancellor, there were deans of various subareas. MRC came under the jurisdiction of the Dean of Letters and Science, a key member of the administration who had significant influence in budgetary and personnel matters. MRC was fortunate, for most of its existence, to have little need for direct support from the dean. That changed at the end, however, and then the administration's inability perhaps to understand, and in any case to supply, the kinds of incentives necessary in a competitive situation was a major factor in MRC's demise.

- *The degree to which a sponsoring agency can itself retain freedom of operation over a lengthy period, as opposed to having to accommodate pressures for short-term results at the cost of de-emphasizing a focus on long-run research.*

As noted in the first paragraph of the last section, MRC was initially given a broad mission, covering four somewhat disparate areas. It was left to the director to decide which areas should be emphasized and how pressures for short-term results should be accommodated. The Army Mathematics Steering Committee (AMSC), to which MRC reported twice a year, oversaw this. This body was valuable to both MRC and the Army. It facilitated contacts between MRC and the Army's laboratories and agencies, but it had other responsibilities. It was not a suitable forum for the sponsoring agency to assess the technical quality of research performed by the center. The solution eventually adopted by ARO was to use periodic peer review on technical matters, with proper sensitivity to the interest of the Army's long-term priorities.

There are very interesting questions about the way in which organizational changes in the Army, and more broadly in the nation's system for supporting research, influenced the directions of that research and the length of the time horizon. These extend far beyond the single case of MRC, and although they deserve exploration they are beyond the scope of what we do here.

We conclude by drawing the reader's attention to Appendices 1 and 2, which deal with problems that MRC faced during the Vietnam war (see Chapter 2). Appendix 1, "A Dove's Defense of MRC," by Dean Stephen C. Kleene, is a reasoned response to demands that the University of Wisconsin abandon its position of institutional neutrality with respect to controversial national issues. Professor J. Barkley Rosser discussed a related question in Appendix 2, "The Horns of a Dilemma." The two horns of the dilemma in question were the following: (1) if MRC did not produce work of value to the Army, then the Army could not continue to support it; (2) if MRC did offer work of value to the Army, it would incur the wrath and enmity of a large number of people on the campus who considered it wrong to give support to any Army or defense function. The latter issue was accentuated during the period of the Vietnam War.

There has been a considerable difference in the climate for support of research since the heyday of MRC. Since the 1980s the situation has completely changed. Universities are receptive with open arms to funding from the Army and other Department of Defense agencies (or almost anywhere else), as was illustrated in the last section in the description of AHPCRC and the Department of Defense–sponsored MURI centers.

Appendix 1: A Dove's Defense of MRC

By Stephen C. Kleene[76]

November 5, 1969

The recent criticisms of the Mathematics Research Center, U.S. Army (MRC) are a renewal, with a new focus, of the attempts which have been made to get the University of Wisconsin to abandon its position of institutional neutrality with respect to controversial national issues. All of this ground was argued over in 1967–68 when the University declined to accede to demands (1) that the University take a stand *as an institution* against the Vietnam war, and (2) that it remove from the campus either all defense-related recruiting (Dow, armed services, etc.) or else (to avoid the legal and practical difficulties of discriminating between different bona fide employers) all recruiting. Not only the faculty and administration, but also in the case of (2) the students (by a vote of 5537 to 2357) when the issue was put to them in a referendum on April 4, 1968, decisively turned down these demands. Institutional neutrality cannot mean that no research should be performed, and no services be provided, that contribute support to one side or the other on controversial issues. It has to mean that the University as an institution is not to select one side as alone to receive the benefit of access to its resources while they are denied to the other.

Without going fully into the reasons for institutional neutrality, consider just this one point. If a public university should abandon neutrality, then (at least in the long run) the government which nurtures the university (funds it, and adopts the statutes and appoints the regents that govern it) could not be expected to keep its hands off the university on political issues. The government of the moment could be expected to oblige the university as an institution to be on its side on each major controversial issue.

* * *

Individuals of any group—of the faculty, or of the student body—should be careful to consider the broad principles involved in any issue, and not to be less firm against a demand contrary to their principles just because the sacrifice required does not touch them. It can hardly have escaped notice that the drive against interviewing on campus was strongest in segments of the student body where interviews are generally less important than, for example, to students in the College of Engineering. No one is obliged to interview with a company he does not approve of; the issued was whether his non-approval should justify denying to others the right to interview. A faculty member or student should be careful not to adopt the position that he will protest against the military to the last dollar of someone else's research contract or fellowship (MRC does award fellowships)—unless, searching his conscience, he can say he would do the same if it were his own.

* * *

The decisions that led to the progressive involvement of the United States in Vietnam, or the decision to build an ABM system and to test MIRV (which seem now to be on the way to being implemented) are not decisions by the Department of Defense (DOD) as a professional service, but by the Executive and the Congress. The professionals provide technical knowledge and advice, and carry out policy; not they, but our elected and appointed officials and representatives (the President with his cabinet, and the Congress) make the policy decisions. If a person's conscience requires him to refrain from taking support from those responsible for actions he deplores (Vietnam, ABM, MIRV), consistency would require that he refrain from benefiting from any federal money whatsoever, including fellowship money.

Persons who are employed in the MRC, as permanent or temporary members, are supported to do mathematical research, and (with a minor fraction of their time) to provide the Army with mathematical training and with professional advice on whatever problems it has in which mathematics is or should be used. Participation in the MRC does not make them supporters of the Government (whatever it is at the moment) on matters of policy. Among members of the MRC, there is probably much the same mix with respect to opinions on Vietnam, and on the ABM as among other segments of the University, except of course that persons who are conscientious objectors to participating in any DOD-funded research are not included. Specifically, the staff who are now, or have been, supported by MRC *does* include persons who are bitterly opposed to the involvement of the U.S. in Vietnam and to the present deployment of an ABM system—persons who have publicly dissented from the government's policy in Vietnam, have written letters to the President and to the Secretary of Defense opposing the ABM, who have contributed time and money to the support of political candidates who have opposed the government's Vietnam policy and the ABM. Under the principle of institutional neutrality, attitudes toward Vietnam and the ABM should make no difference to whether the MRC has a place on this campus. But for the purpose of a response to the recent criticisms, we now give the defense of MRC from the point of view of an opponent of the U.S. involvement in Vietnam and of the ABM. We can offer the same argument to supporters, but they will be less likely to need persuasion.

* * *

Lest there be any confusion, let it be emphasized that MRC staff are primarily engaged in unclassified mathematical research, the results of which are openly published; are valuable as mathematics (if mathematics is considered as valuable *per se*); and are applicable in all sorts of directions in our advanced technological society, of which the military direction is only one. Only a very small part of the staff's time is used in consulting on military applications, and this is done off campus. This is why the activity is appropriately carried on at a University rather than at a separate military installation; in particular, the staff required for this research is of the academic type. (This deserves more space; but it has already been said quite fully elsewhere, though it has been misinterpreted and misapplied by some persons to insinuate that all UW personnel who were ever connected with MRC are part of "the military-industrial complex.")

* * *

If one is against what the U.S. recently has done with its military might, and with the Administration's plans for the ABM system—if he regards the latter as an escalation of a self-destructive arms race—a step which, since it will certainly be matched by steps on the other side, and for technical reasons argued by many scientists, cannot increase our ultimate security, and which diverts resources from things that serve *people*—how can he still believe in doing military research? (The price tag on the ABM proposal, which perhaps is a down-payment only, is something of the order of magnitude of $10 *billion*; a year's operation of MRC is one and a fraction *million*.) A quick answer is that this same person 30 years ago considered Hitlerism to be an abominable evil to be stopped at all costs; is thankful that the allied powers rather than Hitler developed microwave radar (without which the Battle of Britain would probably have been lost); and agrees that, when the atom bomb was recognized by scientists to be an actual possibility, it was an unhappy necessity, but a necessity, to be sure that our side got it first. Let those who now advocate boycotting any military research, or even any research which could be used by the DOD, ask themselves whether they think that, granting all the imperfections of the world as it is, a world dominated by Hitler and his political heirs would be better. Remember that research (military or other) takes time. A person who disapproves the present applications of our military power, but believes it was rightly used in 1942–45 and might sometime again be rightfully needed, could not count on the knowledge to keep this nation on a par militarily with others being available when needed if military research were dropped until the need is imminent.

* * *

A possible philosophical and ethical position is that war is the supreme evil, and that it would be better for a nation to offer no resistance to whatsoever power may seek to impose its will by force than to maintain and to use the means of resisting. We can respect that position, while most of us do not accept it ourselves. (Not the least of the difficulties with that position is that it offers the prospect of only temporary surcease from military activity. Often in history such a nation has been taken over by another power, and then sooner or later its citizens have been conscripted to serve the war machine of the conquering power in one capacity or another, sometimes as slave laborers, sometimes even as soldiers.) What we would make clear is that those opposing all military research (not just abstaining from engaging in it themselves) are in effect advocating the aforesaid position: dissolve the DOD, tear down the Pentagon, abandon our armaments and dismiss the armed forces—while the rest of the world is not doing likewise. Those of us who do not accept this position can believe that military research is a misapplication of science, but that it should not be discontinued unilaterally. That a moratorium on military research be negotiated is an event devoutly to be hoped for. But it does not appear to be about to happen. It will be difficult enough, though we are not entirely pessimistic, that limitation and hopefully reduction of armaments—armaments actually stockpiled or deployed—can be negotiated, with safeguards acceptable to all. Whether after this a moratorium on military research—on even the knowledge of what weapons systems can exist, though they are not being produced and stockpiled—could be the next stage must remain for the present in the realm of speculation; the inspection problem would be so much the greater. The economic burden on mankind is primarily in the stockpiling and use of actual armaments, secondarily in their development and testing, and much less in the research.

* * *

These last remarks refer to specifically military research (and with this the involvement of MRC through off-campus consulting is exceedingly small). Now consider basic and applied research in general, only one of the many applications of which may be to weapons. Natural science has not been an unmixed blessing to mankind—along with the good has come the evil of its misuse for destruction by the military, and the undesirable side effects of industrial developments based on it—the pollution and general degradation of so much of our environment. Mankind might be better off if it could have remained in the pastoral age. The view is heard that one should hence have a moratorium on science (or on large areas of science). But it is too late. We are *in* the scientific age. Science has given us a technology, which supports population levels (which we have, even if we deplore them), and for the more fortunate a standard of living, which could not be maintained in a pre-scientific economy. In this respect, history cannot be turned back; nor will revulsion on the part of some against so much of what science has brought stop others from continuing the pursuit of science. No nation can remain in the forefront in a highly competitive world (even if military means of competition could be removed) if it does not keep up its science.

And though our technological and economic power has the possibility of being used either for good or for evil in the world, it is very hard to argue that we or the world would be better off if there were less of it. Rather than to argue for its diminution or the impeding of its further growth, one had better campaign (if he believes it misused in some respects) for its better use.

* * *

Anyone who would now seriously oppose scientific research and development should reflect that consistency in this position would have required him, had he lived at an earlier age, to have opposed the invention and introduction of the use of fire and of the wheel, and the domestication of horses and of elephants. All four of these discoveries have had military applications; and the first two of these are still of military importance. Had there been a prophet in man's prehistory who foresaw the evil applications of fire and the wheel, how likely is it that he could have stopped mankind from acquiring and using the knowledge of them?

A modern exanple is the discovery of practical applications of electricity.

* * *

We should abjure moral absolutism—which itself throughout history has been a source of great evils. Your conscience may tell you not to do any research that can have military applications. But others, in equally good conscience, can draw other conclusions.

It is no more appropriate that someone with your point of view on Vietnam and the ABM should attempt to forbid a University scientist to do research which can have military applications and to advise the government on defense than that University scientists should be forbidden as citizens to work against Vietnam and the ABM.

* * *

The actuality is that we do have a defense establishment, that this will not simply disappear on our initiative, that our substance (our tax dollars) is paying for it, and that our youth who have been drafted and our citizens who are its professionals are carrying out the missions which political decisions have given it. How would the situation be improved, then, by attempting to deny to the DOD access to the best technical advice it can get on the uses of mathematics in its research, development, procurement and operations?

Is it not better that the Army should have contact with people in the universities whose primary commitment is to education, scholarship and the enlargement of knowledge, than that it should become even more than it is now a self-circumscribed entity? The MRC is not a think tank devoted to thinking up and promoting grandiose new weapons systems and developments to consume our resources and escalate the arms race. To the extent (very small) that people in the MRC have an opportunity through their contacts with the Army to influence policy, we would think their influence would be on the side of restraint.

In summary, it is a consistent position for men of good will that, *so long as* the nation through its elected political leaders has assigned the Army tasks to carry out, those of us who can should help the Army to receive the best scientific advice it can get on how to carry out as successfully and economically as possible whatever tasks it has now or might have in the future, at the very same time that we may be attempting, through political channels, including appeals to public opinion, to alter the present assignments of tasks—to change the nation's course towards disengagement and deescalation in military operations and developments.

Appendix 2: The Horns of a Dilemma

By J. Barkley Rosser[77]

March 19, 1971

Basically, the two horns of the dilemma in question are: (1) if MRC does not do work of value to the Army, then the Army cannot continue to support it: (2) if MRC does do work of value to the Army it will incur the enmity of an appreciable number of people who consider it wrong to give any sort of support to any Army function whatever.

A key point is what is meant by the word "value." The interpretation of the word by the Army is sufficiently enlightened that a considerable fraction of all mathematicians are "guilty" of doing work of value to the Army, whoever is supporting them. Indeed, many critics of MRC will not believe that anything so enlightened is permissible, and accuse MRC of clandestinely doing quite nefarious things in order to earn its support. Even if there were not disagreement as to the meaning of the word "value," the dilemma still has the two horns: no support is given unless "value" is received; consequently, support is incontrovertible evidence that "value" is being given (not that this has ever been denied; MRC does give value, which is why support is maintained).

If one refuses to be distracted by minor clauses of the MRC contract, one sees that basically it requires two things: (1) that U.W. establish a top quality center for research in applied mathematics; (2) that there be a pipeline from this center to the Army laboratories. Many features of the MRC operation are distinctive compared to the usual small individual DOD contract. Most of these features are advantageous, but heavy criticism has been directed at the requirement for a pipeline. Such a requirement is not unusual for the larger centers or institutes contracted for with U.W. As noted by Vice-President Percy, about half of these have a requirement for consulting. Some, such as the Institute for Research on Poverty, set stronger consulting requirements and allow less discretion as to the choice of research than is the case for MRC. Nor is the need for security clearance unique; the Space Science Center has a similar requirement.

The existence of a first class research center in applied mathematics is certainly of general value to the Army. It would be pleased if such a center were supported by NSF, the Ford Foundation, or whomever. However, there are jealous guardians of the Army dollars, both in the Army itself and in Congress, and a more specific case for value to the Army needs to be made to get Army support in the amount needed for MRC. The existence of the pipeline suffices. Through it there is a considerable flow of reports on new mathematical developments, etc., though not very much in the way of specific consulting. Nevertheless, the fact that there is a pipeline, and that it does provide for a flow of specific consulting is a key point. Members of the Army Mathematics Steering Committee (AMSC), such as the Chief Scientist for the Army Engineer Corps, who learn of people in their laboratories who are having mathematical difficulties, have only to pick up the phone to have them put in touch with outstanding mathematicians from all over the world. There is no question that members of AMSC value this. Some would like a larger pipeline, but there is general satisfaction. If the budget of MRC runs into trouble at a higher level in the Army, or before a congressional committee, distinguished Army scientists from AMSC can be called upon to certify the value of MRC to the Army.

Turning to the other horn of the dilemma, since the pipeline is a key factor in justifying the financial support of MRC, those who wish to insure that MRC give no value to the Army try to get it abolished. "No consulting requirement," and "No security clearance requirement," is the cry. Suggestions have been made for modifications in the nature of the MRC help to the Army. Some are minor, being designed so as to sidestep some catchword of the critics of MRC. Clearly this is only a temporary expedient; when the critics notice that MRC still gets Army support, they will be back with more demands. Merely changing the shape of a pipeline will not satisfy someone who is demanding its total dismantling.

For that matter, suppose one could find some quite different arrangement that the Army would consider of equal value. It would do no good to remove the pipeline and set up this other arrangement. What the critics are demanding is that the MRC cease giving value to the Army, and their test of whether MRC is giving value is whether the Army continues to support MRC.

It has been proposed that security clearances be abrogated for the permanent members of the MRC staff. In effect, this would cut off the pipeline outside the doors of Army laboratories where classified work is done, which is most of them. In the fall of 1969 the AMSC considered this proposal quite seriously, and in the fall of 1970 the AMSC considered it even more seriously; both times it declared the proposal too drastic. It has been proposed alternatively that various individual members of MRC voluntarily set up individual consulting contracts with the Army. The pipeline from MRC would be cut off outside the doors, but separate taps inside the doors would be supplied by individuals. Imagine a member of the AMSC trying to defend the value of MRC before a congressional subcommittee. He would say how valuable the consulting agreements of the individuals are. He then would be asked why MRC need be supported; close it down and let him keep the consulting agreements with the individuals.

We are back to the basic dilemma. To justify Army support to MRC, value must be given to the Army by MRC itself; hence the continuance of this support is proof that MRC itself is giving value.

There are a number of options which it might be possible to accomplish which could resolve or abate the dilemma. Presumably a complete resolution would result if other than Army support could be found for MRC. Efforts are being made, but there seem no hopes of success soon on more than a very minor scale.

Another option is to wave a magic wand, after which the Army could give support without having to get value in return. Those who refuse to believe that the present services of MRC are of value, and who consequently attribute nefarious clandestine activities to MRC, would continue such attribution. However, certainly many people would feel happier about the situation. To be supported to do mathematical research quite without obligation would indeed be a delightful utopia. Some have suggested that, rather than a magic wand, it will suffice if the President of U.W. would wave a big club at the Army. However, it should be recalled that giving value as part of a contract lies within the present guidelines set by U.W. For the President to wield a club, either: (1) the guidelines need to be revised, which would require changing the contracts of many centers and institutes at U.W.; (2) special guidelines, applicable only to the Army or to DOD, would have to be set up.

Yet another option, though uncomfortable, is to continue to sit on the horns of the dilemma but to attempt to dissuade as many as possible of the critics of MRC. Some critics emotionally and irrationally transfer their dislike of the war in Vietnam to an antipathy for MRC, because the Army is primarily involved with both. Only termination of the war in Vietnam will dissuade these critics; perhaps not even that. To the rational person there are many points that can be made. The research done at MRC is of high quality, both mathematically and morally. The method of administration of MRC gives many advantages over the small individual DOD grants. The major disadvantage falls well within the U.W. guidelines; insofar as it deviates from the administration of a small grant it is not out of line with many of the centers and institutes attached to U.W. And so on.

Index

Notes

[1] *Science, the Endless Frontier: A Report to the President,* by Vannevar Bush. Washington: U.S. Government Printing Office, 1945
[2] *I.J. Schoenberg: Selected Papers,* Vol. 1, edited by Carl de Boor. Boston: Birkhäuser, 1988
[3] MRC Semi-Annual Report, 25 October 1956, p. 1
[4] *The Mathematician, The Collected Works of John Von Neumann,* Vol. 1, p. 9, edited by A. H. Taub. New York: Macmillan, 1961
[5] MRC Annual Report for 1969, pp. 44–45
[6] MRC Annual Report for 1968
[7] MRC Annual Report for 1969, p. 20
[8] MRC Annual Report for 1968, p. 24
[9] Ibid., pp. 21–22
[10] MRC Annual Report for 1969, pp. 25–26
[11] MRC Semi-Annual Report, 17 April 1970, p. 12
[12] MRC Semi-Annual Reports: 17 April 1970 (pp. 9–10), 2 November 1970 (pp. 9–10)
[13] MRC Semi-Annual Report, 27 April 1971, p. 2
[14] MRC Semi-Annual Report, 1 April 1974, p. 2
[15] *The AMRC Papers,* by Science for the People, Madison Wisconsin Collective (306 N. Brooks St., Madison, WI 53715), 1973
[16] Personal recollection of Dr. Jagdish Chandra, former chairman of the Army Mathematics Steering Committee, 1999
[17] *The AMRC Papers, op. cit.,* p. 103
[18] MRC Semi-Annual Report, 24 April 1968, p. 2
[19] MRC Semi-Annual Report, 5 November 1971, p. 1
[20] MRC Semi-Annual Reports: 3 May 1972, p. 13 and 20 October 1972, p. 16.
[21] *Renewing U.S. Mathematics: Critical Resource for the Future,* Report of the Ad Hoc Committee on Resources for the Mathematical Sciences, The Commission on Physical Sciences, Mathematics, and Resources, National Research Council, Washington, DC: National Academy Press, 1984, p. 5
[22] Ibid., p 6
[23] MRC Semi-Annual Report, 1 April 1974, p. 25
[24] MRC Semi-Annual Report, 3 May 1972, p. 14
[25] MRC Semi-Annual Report, 5 November 1971, p. 15
[26] MRC Semi-Annual Report, 3 May 1972, p. 13
[27] MRC Semi-Annual Report, 18 April 1973, p. 15
[28] MRC Semi-Annual Report, 17 October 1977, p. 19
[29] MRC Semi-Annual Report, 26 March 1976, pp. 9–10
[30] MRC Semi-Annual Report, 11 October 1976, p. 8
[31] MRC Semi-Annual Report, 26 March 1976, p. 10
[32] MRC Semi-Annual Report, 17 October 1977, p. 10
[33] MRC Semi-Annual Report, 10 May 1978, p. 9
[34] MRC Semi-Annual Report, 11 April 1980, p. 1
[35] MRC Semi-Annual Report, 4 October 1978, p. 8
[36] Letter, R. Creighton Buck to H. Edwin Young, 10 July 1975
[37] MRC Semi-Annual Report, 1 April 1981, p. 32
[38] MRC Semi-Annual Report, 21 March 1977, p. 5
[39] MRC Semi-Annual Report, 10 May 1978, p. 8
[40] MRC Semi-Annual Report, 11 April 1980, p. 1
[41] Letter, Jagdish Chandra to John A. Nohel, 13 October 1981; in University of Wisconsin-Madison Archives

[42] MRC Semi-Annual Report, 7 April 1983, cover letter
[43] MRC Semi-Annual Report, 9 April 1984, p. 10
[44] ARO Solicitation Notice DAAG29-84-R-0001, 1 September 1984
[45] MRC Semi-Annual Report, 11 October 1984, p. 7
[46] MRC Semi-Annual Report, 4 April 1985, p. 7
[47] Interview with Irving Shain by Stephen M. Robinson, 30 May 2000
[48] Interview with E. David Cronon by Stephen M. Robinson, 25 May 2000
[49] Ibid.
[50] Personal recollection of Dr. Jagdish Chandra, then chair of the site visit committee
[51] MRC Semi-Annual Report, 20 May 1986, p. 5
[52] MRC Semi-Annual Report, 25 October 1956, p. 1
[53] TSRs 567, 694, 1378, 1396, 1546, 2215
[54] TSRs 554, 723, 852, 1147
[55] M. G. Crandall and P. L. Lions, "Viscosity solutions of Hamilton-Jacobi equations," *Trans. Amer. Math. Soc.* **277** (1984) 1–42.
[56] W. H. Fleming and H. M. Soner, *Controlled Markov Processes and Viscosity Solutions.* Berlin: Springer-Verlag, 1991
[57] M. G. Crandall, L. C. Evans, and P. L. Lions, "Some properties of viscosity solutions of Hamilton–Jacobi equations," *Trans. Amer. Math. Soc.* **282** (1984) 487–502
[58] M. G. Crandall and P. L. Lions, "Two approximations of solutions of Hamilton-Jacobi equations," *Math. Comp.* **43** (1984) 1–19
[59] M. G. Crandall, H. Ishii, and P. L. Lions, "A user's guide to viscosity solutions," *Bull. Amer. Math. Soc.* **27** (1992) 1–67
[60] Ibid.
[61] O. L. Mangasarian and S. Fromovitz, "The Fritz John necessary optimality conditions in the presence of equality and inequality constraints," *J. Math. Anal. Appl.* **17** (1967) 37–47
[62] S. M. Robinson, "Normed convex processes," *Technical Summary Report No. 1135,* Mathematics Research Center, University of Wisconsin, 1971; revised version published in *Trans. Amer. Math. Soc.* **174** (1972) 127–140
[63] S. M. Robinson, "An inverse-function theorem for a class of multivalued functions," *Technical Summary Report No. 1284,* Mathematics Research Center, University of Wisconsin-Madison, 1972; revised version published in *Proc. Amer. Math. Soc.* **41** (1973) 211–218
[64] S. M. Robinson, "Perturbations in finite-dimensional systems of linear inequalities and equations," *Technical Summary Report No. 1357,* Mathematics Research Center, University of Wisconsin-Madison, 1973; substantially revised version published as "Stability theory for systems of inequalities, Part I: Linear systems," *SIAM J. Numer. Anal.* **12** (1975) 754–769
[65] S. M. Robinson, "Stability theory for systems of inequalities, Part II: Nonlinear systems," *Technical Summary Report No. 1388,* Mathematics Research Center, University of Wisconsin-Madison, 1974; revised version published in *SIAM J. Numer. Anal.* **13** (1976) 497–513
[66] E. H. Zarantonello, "Solving functional equations by contractive averaging," *Technical Summary Report No. 160,* Mathematics Research Center, University of Wisconsin, 1960
[67] S. M. Robinson, "Strongly regular generalized equations," *Technical Summary Report No. 1877,* Mathematics Research Center, University of Wisconsin-Madison, 1978; revised version published in *Math. Oper. Res.* **5** (1980) 43–62
[68] L. V. Kantorovich, "Functional analysis and applied mathematics," *Uspekhi Matem. Nauk* **3** (1948) No. 6
[69] P. M. Anselone, *Collectively Compact Operator Approximation Theory.* Englewood Cliffs, NJ: Prentice-Hall, 1971
[70] B. Noble, *Error Analysis of Collocation Methods for Solving Fredholm Integral Equations.* New York: Academic Press, 1973
[71] P. Linz, *Theoretical Numerical Analysis.* New York: Wiley, 1979
[72] B. Noble and M. J. Sewell, "On dual extremum principles in applied mathematics," *J. Inst. Math. Appl.* **9** (1972) 123–193

[73] A. Ben-Israel and T. N. E. Greville, *Generalized Inverses: Theory and Applications.* New York: Wiley, 1974; republished by Krieger, Huntington, NY, 1980

[74] M. Z. Nashed, *Generalized Inverses and Applications.* New York: Academic Press, 1976

[75] F. V. Atkinson, *Discrete and Continuous Boundary Problems,* Academic Press, New York, 1964

[76] Stephen C. Kleene (1909–1994), a noted logician and winner of the National Medal of Science, served on the mathematics faculty at Madison from 1935 until his retirement in 1979. Two of his numerous administrative contributions were his service as acting director of MRC from 1966 to 1967 and as dean of the College of Letters and Science from 1969 to 1974.

[77] This apparently unpublished typescript was prepared by J. Barkley Rosser in 1971, perhaps as an aid in his consideration of the policy issues it describes. Copies were in informal circulation within MRC at that time.

1

2

HEADQUARTERS
AIR MOBILITY R&D LABORATORY
U.S. ARMY

SUMMERS
WELCOME TO
AMRDL

6

7

9

10

11

12